AF697456

The Poetry Company:
Memoirs of A Chicken Catcher

Joe Cobb Crawford

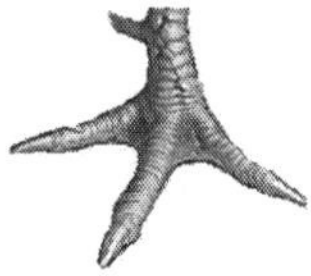

Waldenhouse Publishers, Inc.
Walden, Tennessee

The Poetry Company: Memoirs of A Chicken Catcher

Copyright © 2011 Joe Cobb Crawford. All rights reserved.

Cover photographs by Lester Eugene Crawford
and Carl Franklin Crawford.
Poultry Park Chicken Monument photo by Megan Whitfield.
Manuscript editing by John A. Shivers.
Cover design by Heather Rogers.
Type and layout by Karen P. Stone.
ISBN 978-1-935186-23-6
www.crawfordpoetrycompany.com

Waldenhouse Publishers, Inc.
Walden, Tennessee

Author's Note

This book is a memoir. As such, its contents are of a subjective and artistic nature. I have strived to portray reality as accurately as my memories of fifty years ago have allowed. The availability and accuracies of history itself may also have clouded my representations of these past events. No journals were kept of the stories' events. Photography, though available, was cost-prohibitive to most of the characters included in this writing. The book is about influence. In particular, how precocious youths are influenced by their peers, their family, and other unlikely people and environments. The idea that God gives to each soul a free will to choose whom they will be influenced by is proffered. The intent of this book was not to deliberately vilify or make a hero of any character or archetype. My main goal was like that of the tortoise who found himself with eyes wide open and legs dangling from atop a fence post. He knew he was in an elevated position, and he knew he hadn't gotten there by his own efforts alone. Like the tortoise, I simply was seeking to know who helped me get to my present elevated location in life.

Poultry Park Chicken Monument. Photo by Megan Whitfield.

Dedication

The monument pictured on the facing page honors the champions, the businessmen and the politicians of Hall County, Georgia, who were influential in bringing the chicken industry to the state. Their names and their contributions are forever engraved in bronze at the base of the commemorative memorial. This book is not about them, however. Their names and successes, and their wealth, are well known throughout Georgia, if not the entire world. Instead, this is the story of another group of lesser-recognized, but equally-influential people known collectively as "poultry workers." The "never give up" attitude of these behind-the-scenes individuals influenced and directed my life. It was only after I completed the book's manuscript, that I truly realized the depth of their contribution. This book is dedicated to them: My Fellow Chicken Catchers.

Joe Cobb Crawford
"Little Joe"

Table of Contents

Prologue

Chicken Catchers were a rough lot from a rough environment, namely The Great Copper Basin of Tennessee. The GCB's land, streams and mineral resources had been raped. Progress was the accepted justification for the forced aggression, and people of all walks of life shaped the history of the alledged progress. Their stories are filled with both ironies and anomalies of human behavior. Many had never wanted progress. Others yearned for a way out of their poverty, and progress promised just such an escape. Still others were stranded, having no ticket to board the progress train, but also having no means to return to yesteryear. The latter was the plight of the Chicken Catcher. Their stories are shared in this book.

Mandated restoration of the GCB's land and streams continues today. Not unlike the restoration of human life. Not unlike the memories of a Chicken Catcher.

Acknowledgments

A memoir is an acknowledgment, in a sense, of those people who played an integral part in shaping the writer's life. To call out additional recognition may seem redundant, but there are those individuals who, without their materials, information, and encouragement, this book would neither have been started nor completed. These people are recognized here with humble appreciation.

The front and back book cover are from pictures provided by my two brothers, respectively Lester Eugene Crawford (General Gene) and Carl Franklin Crawford (Pete). The excellent picture of the Poultry Park Chicken Monument was provided by Megan Whitfield, a student at North Georgia State University, and now a beginning teacher in Flowery Branch, Georgia.

The original idea to write a memoir was inspired by my brother Jimmy Dan Crawford (Jim), a retired physics teacher from Dekalb County, Georgia. As an assignment for our family reunion, Jim asked each member to write a "home on the farm" remembrance short story. Chapter 8, "Dog Gone Hogs" was my contribution to that memorable event.

Editorial and proof-reading assistance were provided by three people who deserve to be cited: my wife, Susan, sacrificed tireless hours to help me perfect the words: Charleen Cline, my Friend in Penn, who always took time to listen and to help: Richard Arp, an old friend from high school, without whose help I would have failed Senior English. Rick gave me great advice: "Write the book as best you can, even if you don't try to sell it." Simple words, but oh, so valuable.

Last, but most importantly, I am indebted to Herbert M. Olnick, M.D., the late husband of my sister Doris. His secret encouragement and critiques of my first scribblings kept me alive in the word-smithing game. Herb was the most multi-talented man I've ever known. He was a friend, a mensch, and a true Renaissance man.

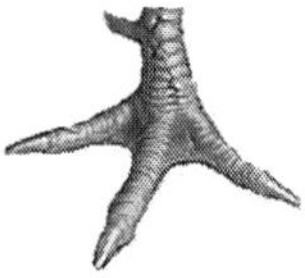

Dictionary of Chicken Catcher Language

Author's Note: Just as today's occupations, such as technology and medicine, have their own unique language and vernacular, the chicken-catching work of my boyhood possessed its own distinct code and way of speaking. This is a list of those terms, which are used throughout the book.

Ass Slap – A chicken catcher tradition. Crate residing catchers slap the head or the buttocks of another catcher as he enters or exits the crate.

B-42 – White leghorn chicken. Small in size, white in color, lightweight bird known to catchers for their rush to flight. Known also for their ability to impart an intense level of stinging pain when they either flog or peck a chicken catcher. If kept in a separate cage, they would have a sight barrier bar strapped to their nose. These barriers made the bird look as though it was wearing glasses. The barrier prevented them from seeing their eggs and attempting to crack and eat those eggs.

Bennies – An amphetamine taken by truck drivers and some chicken catchers. It was known for its stimulating and euphoric effect. Because of the pill's form, it was called "footballs" by the chicken catchers. Truck-Stops were the pill's main distribution source. Side effects of prolonged use included paranoia and schizophrenia.

Bob-Tail – The tractor part of a tractor trailer is disconnected from the trailer part by pulling the fifth wheel pin. The trailer is parked and the tractor can then be driven.

Bomb Tokyo – Have a bowel movement.

Break-down – Remove two and one half ricts of chicken coops (normally about 100 coops) from the front of the chicken truck. Use some to build a set of steps from ground level to about two feet from the level of the bed of the chicken truck.

Broiler – A nine week old chicken usually white in color, and weighing between three to five pounds. Curious and passive in nature.

Bump the Bones – Roll a pair of gambling dice. The dice were at one time made from bones.

Chicken Hook – A steel rod about six feet in length with a wooden handle on one end and an acute angle hook on the other end that is used to catch strays. Also, used to cold cock roosters.

Cold Cock –Strike a chicken on the head with a chicken hook or a long stick, thus making him unconscious, enabling a large or flighty chicken to be apprehended.

Cock Sucker – Prison slang; term used by a chicken catcher to allude to an adversary's homosexual character.

Cowboy – A non-chicken catcher; usually a tourist wearing Bermuda shorts and flip flops; sometimes used in sarcasm to refer to a new chicken catcher.

Crate – A vehicle; usually the panel truck in which the chicken catcher is being transported to the Feather-Nest.

Dead Man's Hand – In the card game of Poker, a hand containing two black Aces, the ace of spades and the ace of clubs. This is the hand held by "Wild Bill" Hickock when he was killed.

Down Below the Mountains – a reference to the northern part of Polk County, Tennessee; usually used to refer to Benton or Cleveland, Tennessee.

Drive Up – Chickens in a large house are coerced to a small containment portion of the house. This is usually performed by first removing all obstacles that might impede the chicken's free movement. Loud, frightening noises made by chicken catchers are used to encourage the chicken's movement to the containment area.

Feather-Nest – A building where the chickens to be caught reside.

Frell hell – To hit or kick hard and repeatedly with intentions of destroying or causing damage.

Greenhorn – A person who is relatively new to the profession of chicken catcher.

Home Runs – A cheap and smelly brand of cigarettes.

Hump – A Jack in a deck of poker cards; original reference is to Jack Hughes, a catcher who had a disfigured and "humped" back.

Keds – A style of tennis shoes worn by boys in the 50s and 60s. They fit high on the ankles, were black with white soles. White strings were used to lace them up. A white spot covered both the inside and outside of the ankle.

Liver Lip – Prison Slang; Demeaning term used to refer to a Negro.

P. A. – Prince Albert; Manufacturer of loose tobacco. The tobacco was placed in a white cigarette leaf and rolled to form a self-made rolled cigarette.

Pavement Princess – A prostitute that catered exclusively to truck drivers.

Payne Gap – A truck stop located on the west outskirts of Cleveland, Tennessee on highway 64.

Poetry Company – A business enterprise involved in handling live poultry; usually chickens.

Punk – Prison Slang; Demeaning term used by a prisoner belittling another prisoner.

Queeney – Prison slang; a catcher who is very young; physically weak; small in stature or having a non-masculine physique.

Rhode Island Reds – A chicken of medium size, dark reddish-brown in color. Good producer of hatching eggs whose shell is light brown in color. A feisty bird known for their quickness at pecking a person who irritates them when they are on their nest.

Rict – A row of columns of stacked chicken coops; usually consisting of four in line columns, each column having ten coops stacked one on top of each other, for a total of 40 coops. This term does not appear to be widely used today, but was standard during the era this book represents.

See a Dog about a Man – Gone to the toilet or to what ever other facilities were available.

Stranded – To the chicken catcher the word meant being left alone without resources needed to do his job or to get home if he had been kicked out from the crate. To the poultry industry, the word related to broiler price volatility: when broiler prices dropped precipitously, but the chicken feed that had been required to raise the broiler had not, the grower was faced with selling his broilers at a loss or holding onto the chickens until the price went up. The under-priced and usually under-fed broiler was said to be "stranded."

Straw Boss – A senior chicken catcher by either age or by years of experience.

Strays – Chickens that have escaped and are running loose around the chicken truck during the chicken loading process.

The Show – The only operating indoor movie theater in the Copper Basin area. It was located in McCaysville, Georgia. A second theater in Copperhill, The Dora-Dale, shut down before 1962. The Dora-Dale was named after Dorothy Strauss, the gaudy glamour missus of a prominent doctor of the Great Copper Basin area.

Time Bait – Prison Slang; A pretty young girl. If an older man was caught dating her, by statutes in some states, he could be sentenced to prison.

White Rock – A chicken, large in size, white in color. They are extremely heavy in weight and slow in movement. Normally their main purpose is to produce hatching eggs that are gathered and taken to a hatchery. Baby chicks produced from the hatched eggs are taken to broiler houses where they are quickly grown off and taken to the processing plant.

1: John Arrah

As a kid traveling at night in a smoke filled panel truck, I recall listening to John R. In the early morning hours, his gravelly, thick voice ruled the radio airways in north Georgia, Tennessee, western North Carolina, and northern Alabama. The wolf-man (Wolf-man Jack) or The Grand Ole Opry would also sometimes sneak in. Our AM radio, usually tuned to John R., and a perpetual poker game kept us all entertained. Before midnight we listened to Tommy Jet's "Jet Fly" program out of Chattanooga. But it seemed that John R. was with us most of the time in that old panel truck. Somehow radio waves from his station, WLAC in Nashville, reached every twisted country road, no matter how buried we were in the folds of those quiet majestic mountains. We were all chicken catchers, and I was the youngest, and John R. was our listening pleasure of choice.

John R. pronounced his name as "John Arrah." We assumed he was a black DJ. The lead-in jingle to songs he played went…

"Hey, John Arrah,
What's you gonna do?
Come on John Arrah,
Play us some… rhythm and blues."

Lyrics heard in the songs played by John R. were slow and clear, and usually repetitious, but they spoke to the hearts of my chicken catcher buddies, a rough crowd of ex-cons, drunks, troubled youths, and other misfits. I was too young to understand fully the meaning of the lyrics, but I sure dug the beat. Later in the sixties I learned the genre was called soul music. The ones who preferred country music despised it and called it N***** Music. I was also shocked later to learn that John R. was actually a white man. Back then, like most white southerners from "whites only" towns, we were willfully ignorant of black people. We listened to

John R's songs in the early morning hours of the late 50's and forgot about our destiny. We were made to feel better about our dirty, smelly job of loading a chicken truck with squawking chickens before sun-up. Thank you, John Arrah, WLAC and Randy's Record Shop of Gallatin, Tennessee, for telling us, with your music, what our closed culture and words failed to!

Joe Cobb Crawford
Chicken Catcher Name: "Little Joe"

2. Wager at The Miami

"What's to come is still unsure:
In delay there lies no plenty,
Then come kiss me, Sweet-and-twenty,
Youth's a stuff will not endure."
William Shakespeare

It was deep summer, evening was coming, and a sweltering blue haze draped the stark mountain background. Fittingly named, they were The Great Smoky Mountains, the place where the Cherokees had escaped to avoid their surrender to the "Trail of Tears" march. The Cherokee Nation Reserve was nearby, but a few hold-outs chose to live near a remote western North Carolina town called Murphy. "Way over yonder" was how most people referred to Murphy. It wasn't really the end of the world, but you felt you could probably see it from there. At the entrance to a cluttered string of drab merchant shops was an architectural anomaly called "The Miami" – a truck stop located on Highway 64. Its large, multi-colored neon sign prominently placed near the highway's edge was a visual enigma. Flaming pink flamingo birds, green palm trees, and a fish in flight were all displayed in glowing neon. All in all, an unforgettable sight. Inside, the motif continued with a large blue marlin mounted on the wall above the cash register. Cherry red Deco-style bar stools lined a long, shiny checker-board patterned counter top. Tablecloths displayed images of exotic swamp birds. The Miami wasn't just bizarre; it was completely out of context. It's over five hundred miles to North Carolina's marshes and ocean coast.

Nonetheless, there it was. A truck driver's tropical paradise located in the middle of nowhere, North Carolina. We chicken catchers also liked to stop in, grab a cold drink, play the juke-box or pin ball machine, before traveling on to the feather nest. A few establishments didn't like us catcher, and would lock their door if they saw the crate roll into their parking lot. They said we stole them blind, dirtied up their place, and made other customers not want to come back. The Miami, on the other hand, liked our business. So, we stopped there.

For those who are already questioning the connection between chickens and a "poetry" company, let me explain now. Otherwise the question is apt to linger in readers' minds for the rest of the book. It was a misunderstanding due to Bill Roach's accent. Bill was a white guy but he spoke, sang and acted much like a black person of that time period. When asked where he worked, Bill Roach once told an inquiring truck driver at a truck-stop that he worked for "The Po'try Company." The truck driver of course heard "Poetry" instead of "Poultry." Bill didn't correct the truck driver's misunderstanding and, since Bill was sort of a poet at heart himself, who really fancied the idea of working for a "Poetry Company," he just let the misunderstanding stand. Some of us chicken catchers found the name to sound funny. Other catchers who truly saw the irony between "Poetry" and a smelly "Poultry" business, also saw dark comedy in the name. We all just liked the name better than the real name, so it stuck.

Tonight we were to meet up with the owner of the chickens. He would lead us from the restaurant back to his feather nest, a place way back in the mountain sticks, and a long ride even after you left the pavement. As planned, he was there, inside, in the midst of the bony bayou birds, waiting on us when the crate rolled in.

As we all entered the truck stop door, a tall, overweight, middle aged man wearing a tan khaki shirt and a baseball cap stepped forward and asked, "Which one of you fellers is the foreman?"

No one answered, but then Glenn Watkins, a ruggedly built and normally reserved teen, who went by "Scuttle Butt," surprised us all when he spoke up. "That's Frank Ray, but he ain't here. He's out there putting air in the tires. The crate had a flat tire on the way up here." Then Glenn sidled up to the man, smiled and said, "But, I'm his 'Right Hand Man'."

Bill Roach, hearing Glenn's gesture, and not wanting to be upstaged by Glenn, quickly hollered out, "Yeah, that's right. Frank wipes his ass with his other good hand." You see, Bill truly believed that he was a born performer, and that the spot light should always be on him. At home no one ever noticed Bill. He was a middle child from a poor family, and they often got evicted for not paying their rent.

He was a short, tough, muscular guy, and his quick temper landed him in a lot of fist-fights. Two types of people irritated and angered Bill: tall people and bossy people. He usually was able to hide his anger toward them, but sometimes his passions would boil over. Unfortunately, his hidden anger consistently boiled over when he became "fall down drunk" and, because of that, he rarely won his fights. His face often looked more colorful than a rainbow and his eyes, if you could see them, looked like a barber's pole. They were usually blood streaked and their whites were rarely clear. More often, they were swollen shut and his bulging cheek bones would beam in glorious shades of purple and orange or yellow.

Bill had a few other passions: he loved music and poetry. He also loved an audience and, when he wasn't beat up too badly or didn't have a hangover, he would quote poetry. Poems he selected were usually from a jail house genre, and their topics were vile and vulgar. And man, could he ever sing! Some of the catchers jokingly called him "Jig-a-Boo" and said he was no blood kin to the Roach family. They'd insist that Bill had been adopted and that his real folks were black and they'd put him out because Bill was an "albino n***** baby." They'd say this because Bill would sing and juke around, and seemed to be on his own private cloud.

The tall guy turned his back to Bill and ignored his wisecrack. He started talking to Glenn, instead. He didn't know it, but he had just made himself an enemy and, if Bill's blood-shot eyes could have killed, a coroner would now be needed at The Miami. The rest of us chicken catchers sauntered in and took a seat in the eating area far away from the truck drivers, but Bill, ever the showman, he pranced on over to the jukebox and dropped a quarter in the slot. He turned then and loudly slapped his hands together, threw back his head, and belted out the words of James Brown that blared from the big record player. His timing was perfect.

Whoa-oa-oa! I feel good, I knew that I would, now
I fee… eel good, I knew that I would, now
So good, so good, I got you
Whoa!

Continuing to sing along, Bill swayed and snapped his fingers and then shuffled on over and sat down at a table.

Marcus Payne, a catcher who had already spent time in both Joliet and Leavenworth penitentiaries, took a seat at a table at the back by the entrance to the bathrooms. Marcus was a small time thief, but he had big ideas and the gears were always turning in his head. He had shifty but mirthful eyes. Though athletic in build, he had a slim chiseled jaw, and a sharp-edged nose that reminded you of a predatory bird. Marcus always seemed overdressed for the dirty job of catching chickens, and he would wear patent leather slippers to work in. Royal Crown hair dressing usually drenched his nappy hair and, when styled, the set waves were sheltered by a do rag. The sister of Bob and Bill Roach would one day have his baby. On learning of his fatherhood, Marcus had ruthlessly declared, "If the little bastard is born with one of those damn n***** noses like the Roach's have, I'm gonna' drown him. He needs a sharp nose like mine, so he can cut the wind easy when he's running from the law."

The fate of the unborn child would be better than Marcus's calculating mind could ever imagine. Much better than his own fate. His father had been killed in a mining accident at Burra Burra Mine at Ducktown.

Upon sitting down, Marcus's eyes began scanning the area near the cash register. Together, he and another ex-con named Bill Carter from Turtletown, discussed how they could distract the waitresses long enough to take the money from the till without being seen.

Two tables away a few of the other catchers who had no pen time sat, appearing to kill time. In reality, they too, were plotting a theft, but one of a smaller scale. Their plan was to take the clear glass pie holder and the pie in it that set out on the bar counter. They first had wanted to take the marlin, but saw no way to pull it off. Their second plan would be executed as the crew ducked out of The Miami, headed for the crate and a night of work.

At another table, between these would-be absconders, I sat with two greenhorns. Both were nervous and their feet twitched and their legs jittered under the table. I sat with them because I felt we had something in common, and it wasn't that we knew nothing about chickens. I knew chickens well. My two brothers owned the Crawford Poetry Company and I lived on a farm and had been around chickens most of my life. What we did share in common was fear and anxiety: Could we cut the mustard and gain acceptance from the crew? This was always my biggest concern, because I was the youngest and the smallest catcher in the crew.

Each of the greenhorns wore a white T-shirt with a pack of Camels rolled up in its sleeve. Each also had a heart shaped tattoo on his right bicep. One read "Mom" in the middle. The other had a dagger through the heart's middle with words "Born to Lose" below.

Seeing the Camels, a catcher from the other table asked one of the greenhorns, "Spare me a smoke, greenhorn." Then, he turned to Marcus and said, "Give me a match Time Bait."

Marcus stopped scoping, stared suspiciously at him, and quizzically responded, "A match you say, faggot? Yeah, I got a match for you: My ass – your face. It's a perfect match." With a conniving smile he then tossed him his lighter.

Before the waitress came around, an always upbeat catcher nicknamed, Limpey bounded up to the tables. Limpey was a clean cut boy, a good worker and was now the man of the house. His dad had been killed in a car wreck and Limpey caught chickens to help his mom pay the family's bills. He took his new family role seriously, but he also wanted the crew to like him. Limpey's real name was Jimmy Wimpey, and he got his nickname from W.B. Ray, a senior catcher who had a speech problem. W.B. pronounced many words wrongly and no matter how hard he would try, the word "Limpey" came out when he tried to say "Wimpey." Limpey stuck.

As though he was trying to keep a straight face, Limpey loudly announced, "Did ya'll hear about the bet? Watkins told that big lard-ass guy up there that we'll have all his chickens loaded and be back here to The Miami by daylight in the morning, if he'll buy us breakfast."

From a nearby table came sounds of disgust. "No f#?!/#? way. Hell, we don't even know where the feather nest is yet.

Another added, "No chance in hell that's gonna happen." We got two double axel trucks to load tonight."

Another asked loudly, "What's Scuttle Butt's ante on this bet?"

"Narry a dime," replied Limpey."

"That lard-ass sum-bitch is crazier than he looks" was the last derisive words voiced about the wager. The chatter then subsided and a shapely waitress dressed in a skin-tight dress the same color as the neon flamingo approached our table.

Seeing the tattoo with "Mom" on it, she commented to the catcher, "Love that too! My last husband had one just like it. Where'd you do your time?"

"Treasure Island in Frisco mostly. Got dog drunk one night with some more sailors and done it. Wish I hadn't" he said while placing his sweaty hands on his nervous legs.

"Oh. I'm sorry. What can I get you hon?"

Bodies leaned and necks stretched as they peered around each other to get a glimpse of her. Eyebrows were raised to gaze beyond her hemline's boundary, and lewd whispers about her were overheard as we each placed our order. She couldn't hear them, though; foul language in the presence of a woman would get your ass kicked in a heartbeat. It was against the chicken catcher's code.

Just then two of North Carolina's finest with shinny pistols on their hips entered and sat on the chrome, red-topped bar stools, where they ordered coffee and pie. All talk of thievery amongst the catchers stopped.

We finished our orders without a word and paid our bills, all polite like, then left with Bill Roach snapping his fingers and dancing toward the crate as he sang, Hey, John R, What's you gonna do, Come on John R, Play us some rhythm and blues.

The officers heard the singing and looked out from The Miami, then shook their heads in disbelief. As we got into the crate we heard the chicken owner tell Frank, "Okay, Captain. Follow me to Bloody Bucket."

3. *Big Breezy*

"The great beauty of Poetry is that it makes everything, every place interesting."
John Keats

Bloody Bucket they called it. Once, it had been a Cherokee Indian settlement, but now it was mostly inhabited by poor, mountain dirt farmers, and a few people who kept laying hens. The road from Murphy into Bloody Bucket was like most other roads leading to a chicken house – narrow, curvy, and not designed to handle a sixty-foot, eighteen-wheel chicken truck. A narrow bridge over the Hiwassee River had to be crossed to get to Bloody Bucket.

The chicken house was also long and narrow. Ridged rows of a crusted fusion of chicken dung, wood chips, and ragged feather stubs ran the length of the house. Slogging over these jagged pathways all night with two hands full of flogging white rocks would be tough.

"It's the type of feather nest that'll put muscles in your shit," was how the foreman, Frank Ray sized it up, after entering and making a trip down through the house.

The crate that transported us catchers arrived at the feather nest first. Chicken trucks on mountain roads were often delayed and those shabby roads were often the cause. They were usually narrow and had many short, winding turns. Because of that, a tow truck sometimes had to be used to help them get into these remote locations. The narrow mountain roads made navigating a chicken truck both a slow and dangerous task, and sometimes we catchers had some time to kill. That was the case this night.

As the doors of the crate swung open, a cloud of cigarette smoke rolled out and a slow random popping sound was heard as about six, travel weary guys, their heads ducked, tried to exit. Some rubbed their buttocks. Others straightened their caps or hats as they emerged. The sound heard had been the traditional parting ass slap, a skill requiring both touch and finesse. Too light an ass slap would get you labeled as a Queeney or a faggot by your peers. Too aggressive an ass slap would often result in a fist-fight between the slappee and the slapper. No fights today, though; a hot game of five-card stud poker held everyone's interest.

"A hump and no help on your straight" said the dealer as he adjusted the make-shift coop bottom/card table and continued to deal the cards. Eventually, the remainder of the guys emerged from the crate at the end of the last poker hand.

Noting the absence of a chicken truck and being the apparent game winner, Glenn Watkins, forever to be known as T-Bone after this night, made a suggestion: " Hell, boys, we ain't got no truck to load. Let's play a few more hands." At this suggestion, the chicken coop bottom that had been used as a poker table was retrieved from the truck. While waiting for the card table, T-Bone rolled himself a P.A. and watched the sun starting to sink behind a mountain. The game resumed in a grassy area near the front of the feather nest. The straw boss ordered a couple of non-players to ready the drive up, while others stayed to watch the game.

Full of wanderlust, my two twin buddies, Carlie and Charlie Tankersley and me, drifted off to the back of the chicken house because we wanted to check out the rough terrain. But of more interest to us was the big-ass fan located at the far end of the house. We had never seen anything like it before, and wanted to get a closer look without being observed.

So we walked over the ridged rows of the poultry dung to the other end of the chicken house. The fan was enormous and stood out as a prominent point of destination. Located at least three hundred feet away, the fan's support structure could be seen as well, even as we entered the front of the house. As we walked, Carlie was singing a single verse of a catchy new song… and singing that same verse over and over again.

Elvira. Elvira. My heart's on fire.
For Elvira.

Growing annoyed with the repetition, Charlie cut his eyes at his brother and blurted, "Hey, Elvis! Why don't you shut the f*** up until you learn the rest of that goddamn song?"

"Hey, Cock Sucker, who the hell died at Bandstand and made you Dick Clark? By God I'll sing if I feel like singing," Carlie fired back at his brother, as he snapped his head toward Charlie, his eyes squinted in anger. He was clearly pissed.

We continued walking, now, with the verse permanently burnt on our brains.

The Tankersley twins were a little older and bigger than me. They were tall, wiry and had jet black hair, which they always kept groomed to perfection; Carlie especially. Seems like every nine seconds Carlie's hands were seen combing back his wavy mane. The other catchers didn't like them and called them "the STANKersley Twins," but I always found them fun to be around. They lived on up Mill Creek, up the creek from our farm, up near Wash Wilson's apple orchard, but they didn't farm or grow apples. Their family's house was small and they barely got by on their dad's meager disability check. We were much alike in one way – we all liked having a little money in our pocket. So, we caught chickens to have that money. The twins were always up for the next adventure, and could quickly create one if none presently existed. One strange behavior pattern had always bugged me about these twin brothers, though: they were always pestering each other and arguing between themselves. They cursed each other frequently and, in listening to them, you expected a slugfest to occur at any minute. Some people who had TVs, referred to them as "Heckel and Jeckel," two mischievous crow characters who were the stars of a cartoon program.

When we arrived at the big fan, we were even more awe-struck by this giant air machine. We stood back from it and positioned ourselves near the center of the fan's wide wing span. The blades of the fan were larger than life. They covered almost the entire end of the chicken house and the blades turned so fast that a blur was about all that could be seen. The roar of the rushing air from the fan was refreshing, and we had to lean into its blast to keep our balance. The air was gloriously absent of the stench of chicken manure and it was cool, but we had to shout to each other in order to be heard.

We were still entranced and refreshed by the monster fan's cool breeze, when Carlie shouted, "I wonder if I can put a chicken through that fan?"

Charlie, almost as though he had been considering the same question, doesn't miss a breath and yells back, "Hell, no! You idiot! You can't even see outside through that fan. How you gonna get a chicken through those blades?"

"What do you think, Little Joe? Do you think I can do it?" Carlie asked.

"You damn tootin'. It could be done, I think." But, truth is, I didn't really think he would actually try anything so stupid.

Encouraged by my phony confidence, Carlie quickly shouted back, this time even louder, "I got a dollar and a quarter says I can."

"Put your money where your mouth is then," Charlie shouted back.

The bet was on. I was the money holder. Ten quarters – five in each pocket of my worn Wranglers was entrusted to me for safe keeping.

Carlie made steps to find his test chicken, but Charlie grabbed him, hauled him up short and shouted, "Hell no, you don't. Joe's the one that'll pick the chicken for this goddamn bet!"

Carlie's head bobbed in agreement. Wasting no time, I grabbed the first chicken that walked within snatching range. It was a big rooster and he strutted slowly by us, like he owned the place. He was almost arrogant and seemed to be indignant about us being on his property. I held him high above my head and shouted, "With the power invested in me, I christen thee "Big Boy."

The twins gave me a blank look stare – like the stare a calf gives to a new barn gate. I hoped that one would back down from their bet but they didn't. They both stood stock still and voiced no second thoughts. I feared the unknown outcome of the bet but I tried to not let my fear show. Why had I told Carlie he could pull it off?

Big boy was a credit to his species. He would easily have weighed twenty pounds, and was a peaceful soul that put up no fight to escape from his imminent sacrificial mission.

Carlie took Big Boy from my hands. Then he tightly grasped both of the rooster's legs and began a blazing softball wind up. Big Boy, stunned by this unknown gyrating motion, made no perceptible sound or untoward motion.

Then Carlie released that rooster and, like a rocket, he traveled toward the magnificent fan. But the closer he got, the slower his speed became. Not once did he try to flap his wings and escape, and his trajectory placed him low and outside. He landed, wedged in the bottom right corner of the fan's massive frame.

Immediately, feathers shot towards us like arrows that had been launched from an Indian war party. Pieces of chicken flesh and blood splattered our faces. Spines of feathers lodged in our bare arms and clothes. Big Boy made no sound throughout his execution.

The long strings of overhead light bulbs suddenly began to flicker and flash, and it seemed like a cosmic electrical storm was taking place inside the chicken house. Then, as though an invisible hand of God had grabbed it, that big-ass fan instantly stopped turning. Everything got deathly quiet and beads of sweat instantly popped up on my forehead and my arms, as a knot gripped the pit of my stomach. No air moved and time seemed to stand still. All that was heard was a low humming sound coming from the center of the fan. Even the chickens were silent, and they all raised their heads, as though they, too, sensed something horrendous was about to happen. They were right! In an instant, a cloud of smoke burst forth and it floated upward, followed immediately by flashing electricity. Finally, a shot of fire burst out at us from the center of the big ass fan.

Charlie screamed, "Let's get the hell out of here!"

"Haul ass, Joe," Carlie yelled as he shoved me from where I stood, frozen and mesmerized, still in a state of shock.

"How do we get out?" I shouted back."

"Over there. I see a side door," Carlie cried with panic evident in his voice. "Run!"

Safety was in sight now and almost as if we were yoked in harness, we all crashed through the door that led us outside. Once freed, on the outside now, we still sprinted like scared deer toward the front of the chicken house, where we breathed a little easier. Daylight was almost gone. A few of the catchers who had lost all their money in the poker game were standing at the side of the house, and we saw them, so we slowed to a walk to avoid raising their suspicions.

As we slouched innocently by them, one asked, "What the f*** is going on inside there?"

Charlie, always quick with words came back with, "How the hell should I know? I'm just a damn chicken catcher."

A greenhorn, the one who had the nervous feet at the Miami, was standing with the group, and he asked, "Which one of you mother-f****** is smoking those goddamn Home Runs cigarettes? I tell you one thing, if you son-a-bitches are smoking those damn things when we go home in the morning, I'm gonna throw you and those raunchy ass smokes out the crate. You can take it to the bank. Them damn things gives me a sick headache to smell em."

Bob Roach, a pensive and passive sort of guy, and Bill's older brother, didn't like the threatening tone. With a curt smile playing about his lips, he shot back, "If I had a damn head like yours and it didn't ache, I'd go see a doctor."

The greenhorn looked closely at Bob but said nothing, because he could tell from Bob's composure that he didn't want trouble with this guy. He also suspected that Bob, like several of the catchers, carried a switchblade.

In the fresh air we, too, could smell and taste a peculiar odor. Smoke from the flaming fan motor had stuck to our hair and clothes, but we acknowledged nothing.

Tensions were high among all the guys. Still, the twins and I were now strangely relieved. We were now in a much better place than we had been just five minutes before.

Out of the tense silence another greenhorn nearby was heard to exclaim, "This goddamn Bloody Bucket place here gives me the heebie-jeebies."

Without batting an eye, Charlie took that as his cue to change the subject. "You damn straight, man. Hey, Little Joe, you still holding my money?"

Indiscreetly, I emptied the ten quarters from my pockets and handed Charlie his winnings. The long awaited chicken trucks soon came into sight from around the mountain side. Slowly, like prehistoric monsters, they crept up the dirt road to the front of the dark and now "fan-less" feather nest.

That was when Carlie, started up once again.

Elvira… Elvira. My heart's on fire. For Elvira.
She gonna' scream and holler
Cause I saved up two dollars
We gonna' search and find that preacher man.
Elvira… Elvira…

4. Reckoning Day

"Lo, how men blame the gods! From us, they say, comes evil. But through their perversity, and more than is their due, they meet with sorrow."

Homer: From The Odyssey

The neon sign of The Miami had never looked so heavenly. Here we were, back where we had been only ten hours earlier, and seated at the same tables near the back. This time some things were different.

All the catchers looked spent. One late arriver from the crate surveyed the crew sprawled out around the tables, where some were laying their heads on the tabletops, and announced, "You f*****s look like you been shot at and missed but shit at and hit."

No one disagreed with his wisecrack, because we truthfully knew that we probably did really stink up the place. We all were sweaty and filthy, and we were way beyond being just tired. A fine dusting of chicken manure coated our clothes and our bodies. Sweat streaks lined our dirty faces; our nostrils felt crusted inside. The itching of the crawling chicken mites on our bodies was ignored; we were dog tired and didn't feel like scratching. Every man was feeling the toll of his work and, while we might have been in bad shape, we weren't defeated. It wasn't anything that a good shower and forty winks wouldn't cure.

Seated with me this time was Glenn Watkins and, as the sun came up, he cut into a well-done T-Bone Steak. The rest of us, enjoyed a Honey Bun, a Moon Pie, or some dounuts with our Double-Cola drink. We had done it! By working together, we had met this impossible deadline, and each catcher knew it. Success was a balm and no one begrudged Glenn eating his steak in front of us. But why was T-Bone (alias Glenn Watkins or Scuttle Butt), eating a five dollar T-Bone Steak – more money than any of us had made that night? The answer was, T-Bone ordered last, and when asked what he wanted, he had answered, "I want a T-Bone Steak. I sure as hell don't want any damn eggs for breakfast." The big guy chicken owner at first had protested T-Bone's order, but relented when T-Bone reminded him that no restrictions had been placed on what could be ordered for his Victory Breakfast.

Somehow, when we started to work that night, everything had just clicked. The lights being out in the feather nest was only a minor problem. Head lights from the crate were used to break down the truck, to build steps and to begin loading the chickens. When the lights were restored, we drove the chickens up closer to the truck and corraled them close together with a few coops borrowed from the second truck. Everyone cooperated with each other and, oddly, the usual covert egg hurls and the subsequent threats by a stung and splattered catcher were absent. All the previous tensions were soon channeled into the manpower needed to carry and muscle the heavy hens up to the truck. Even Bill Roach pushed us to work hard that night. He wanted to stick it to that big guy any way he could for his snubbing him. No mention was made by the owner about the lights being out and the big-ass fan not running.

The next day, after sleeping until around four in the afternoon, I walked over to the Crawford Poetry Company office. I knew that Frank Ray would be going there to get another spare tire for the crate, so when I got there, I bought a Coke and some peanuts from the vending machine. I sat down on the office stoop, put my peanuts into the Coke bottle and waited. He had told us the night before that we had a load going in Hanging Dog, North Carolina this evening.

A truck driver raced up the stoop and asked me, "How you doin', Little Joe?" Then he went on into the office. When he came out a few minutes later, on his way to his truck, and as though he had just remembered, he yelled back, "Brownie wants to see you."

My mouth was full of Coke and peanuts, so I made the sign of a circle with my thumb and forefinger and the other fingers standing at attention, and finally yelled back to him, "I'll see you a little later up at Hangin' Dog."

As I entered Brownie's office I could see the color of his normally pale face start to turn red. He got up from behind his desk and walked over to me, got up close in my face and shouted, "What the hell were you assholes trying to prove last night up there at Bloody Bucket?"

I looked at the floor and said nothing.

He continued his tirade. "Did you know that you clowns did over two hundred and fifty dollars worth of damage to that man's chicken house last night? He's mad as hell about it, and I don't blame him. But he's not mad at you guys. He doesn't know you bozos. All he knows is the name he sees on the doors of those chicken trucks, and they say 'Crawford Poultry Company'. That's the name of the company, and your brother Jack and I own it. That makes us responsible."

The office then got as quiet as a church house on Monday morning, and it stayed that way for twenty hours it seemed. I peeked up now and saw that Brownie had begun to cool down a little. His face was still pink, but its color was returning to normal. He walked away, just a couple of steps, then he turned quickly and said in a voice that was suddenly restrained, "The next time you get ready to do something stupid like that, I want you to stop and use your damn head. I can't afford to pay the bill on all you and your twin buddies can tear up when you don't think."

Trying to keep my chin from trembling and still looking at the floor, "I whispered, "Okay." Stunned, confused and hurting I left Brownie's office and walked back out to the stoop. I felt worthless and lower down than a whale turd. I wanted to cry but remembered that crying was not an option for a chicken catcher. Criers are queenies by the catcher's code. Fighting back tears, I retrieved my Coke and peanuts, sat back down on the stoop, and tried hard to think about other things… anything. Luckily, Frank Ray soon rolled up in the crate, pulled it up to the gas pump and he got out. Dashing up the stoop, he called over his shoulder, "Fill it up and get that bad tire out of the back. Take it to the basement garage and get that spare that Jack fixed for me today."

Feeling a need for any kind of redemption, I jumped up and did as he asked. He returned just as I had loaded the spare tire in the crate. We got in and slammed the doors, and headed off to round up the rest of the crew.

Feeling better now, I told myself, "It's going to be a great night! I'm getting to ride shotgun and I won't have to ride in the back of the crate and breathe the gray haze of cigarette smoke." Then I stuck my head out the window and breathed in a few deep breaths of air, as the rough-riding, old panel truck sped off leaving a cloud of dust behind.

5. The Pit

"The whole universe is change, and life itself is but what you deem it."

Marcus Aurelius: From "Meditations"

Far from the chicken house site there was always found a large hole dug into the ground. This mysterious and concealed outpost was called "The Pit" and it entombed deceased chickens. This underground house of death was purposefully placed off the beaten path to assure the safety of farm animals that might accidentally stumble into it. The structure of this massive tomb was simple: A deeply dug hole was covered with wooden timbers. In the middle of the covering was a bottomless barrel that stood up high like a chimney. The barrel's top, seemingly benign, was a hinged lid and it served as the entry point to this chicken grave. What lay on the other side of the barrel lid was a Dante's Purgatory for dead chickens.

No light entered this modern day hell. A well-focused flashlight was needed to view the contents – even at mid day. A long stick or a chicken hook was needed to flip open the lid, whereupon an explosion of escaping green flies would instantly appear. Next, on the underside of the lid, was seen a glinting and crusted formation of slowly creeping ebony bugs. Then it hit you – the smell. A rancid, musty, transient whiff that pierced your nasal cavity right between the eyes and brought on an instant headache.

If one dared to go closer to the opening, he heard a faint buzzing sound – a sound like honey bees at work. But it wasn't honey that was being made. Closer still, one could see the source of the sound – over a million newly-hatched flies randomly testing their wings for the first time. Maggot colonies swelled, oozed and rose up toward you. Every instinct in your body told you to run like hell from The Pit! But, if you could hold your nerve and hold your breath, and hold down your last meal long enough, other hellish glimpses were to be seen down there.

At the edge of the rotting pile of blood and guts and feathers, near the Pit's clay bank walls, could be seen cold and inanimate beady eyes. A beam of light rendered their owners statuaries. These were rats – not mere mice – but gigantic, sneaky creatures with ivory pointed teeth and bloody lips. They were so big they could be mistaken for possums that also entered The Pit. Foxes and other carnivorous rodents sometimes were seen in there as well.

The barrel lid was the only known access to The Pit, but somehow these outside critters managed their way in to this death hole. They were welcome to it.

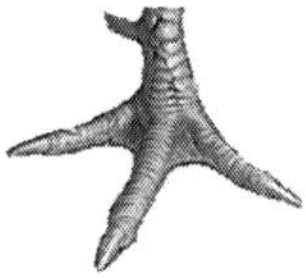

6. *Ham Hocks At Hanging Dog*

"Fair is foul, and foul is fair: Hover through the fog and filthy air"

William Shakespeare: From "Macbeth"

I had lost my shotgun seat. Randal Cearley was the first of the crew roundup and, being a senior catcher, was entitled to the preferred right front passenger seat. This seating priority was never voiced, but was understood by all the catchers. The hard steel lid of a five gallon iced water tank was now my seat. This "up front" icy seat would both numb and freeze your butt at the same time, but it was better than riding in the back of the crate. No smoke up front, just plenty of fresh air.

Randal Cearley was complex. He was clever and seemed to have some education. Sometimes he brought a Zane Grey paperback to read while traveling in the crate. When W.B. Ray, a senior catcher, first saw Randal reading that book, he had commented, " I be damn, Sir-lee! I didn't know ***you*** could read." One of W.B.'s speech problems included adding the "sch" sound to many of his words. Once, a waitress asked W.B. how he wanted his eggs prepared and he replied, "Oh hell, just schri um, just schri um." Unaware that he was answering her in the only way he could, the waitress told W.B. they could only scramble, poach, or fry the eggs. In addition, W.B. was illiterate, but could sign his name on his paycheck.

Though older than most of the catchers, Randal acted youthful, was good humored, and shared their fascination of a good laugh. He could tell great stories; tales that were easy to believe. He never let little things, like the truth for example, get in the way of his storytelling. But his stories always seemed honest and consistent. Poker was not his game. He told me, "It's a game for losers" and, with his thumb, had motioned to the catchers who were playing in the back. His physique was muscular, well-proportioned and like that of a well-conditioned athlete. He was clean, dressed neatly, and always looked as though he had just showered. A pack of Lucky Strikes was usually rolled up in his T-shirt sleeve. His dark blond hair perpetually looked wet, but was immaculately combed and smelled of Vitalis hair tonic. T-Bone once told me that Randal drank Vitalis and put a little on his hair to hide his habit. I didn't want to believe T-Bone. I'd never seen Randal drink anything, but T-Bone would have known – his dad battled the bottle. Randal's face was small and his cheeks sunk in a little, mainly because he had no teeth. Oddly, his face reminded me of those two masks characters that were highlighted on the walls at the picture show in town. Another catcher told me that Randal had once owned a set of false teeth, but went on to say that Randal's teeth were now showcased and up for sale in a pawn shop down there below the mountains. I didn't want to believe that either.

With all catchers in place, the crate was now speeding up Bell Hill Straight toward Hanging Dog. This mile-long, two lane straight road led from Tennessee to the North Carolina line, and hot rods would often drag race there late at night. At the state line was a hodgepodge of small stores, and each one sold tax free North Carolina cigarettes for under two dollars a carton.

A fast moving hotrod passed us and almost collided with an oncoming car. The driver of that other car looked angry as he passed us, and his horn was blaring. Someone from the back of the crate hollered, "Keep it up, you damn speed-demon bastard. Hell ain't half full yet."

These words were wasted on Randal. He heard nothing because Randal was not easily distracted, and he often appeared to be in deep thought. Randal nudged me in the ribs and said with glee evident in his voice, "Hey,Little Joe, did you know we're going back to Indian country again tonight. Yeah we're going up to Hanging Dog. Lots of black bear up there, you know. I've been told that a black bear was what caused that dog to get hung up in the creek up there at Hanging Dog. They say the dog wasn't chasing a deer at all. No, he was chasing a black bear. Smart animals, them black bear. The old sow led that dog into a trap set in the creek. The dog got stuck, couldn't get loose, so he starved and drowned to death. He stayed trapped there in that creek until he rotted and he was finally washed on down stream.

"Better keep an eye out tonight for them black bear. You'll know when they're around though. You can see and smell bear shit and the woods are full of that stinking stuff. Be careful not to step in it. It's as slick as snot on a door knob. So, watch your step outside around the chicken house tonight."

Someone from the back hollered, "Hey Cearley, I think you're full of bear shit."

Again those were wasted words. Randal just kept on talking to me. "Did anyone ever tell you the story about how Indians get their names, Little Joe?"

I shook my head. "No, I don't know that one." As I spoke, I grabbed a quick glance into the back, trying to see who Randal's heckler had been.

"Well, here's how." Randal leaned forward in his seat and placed both hands straight up in the air. Then he started swaying back and forth and a loud recurring chant issued from his mouth. "Hey Ya…ah… Hey Ya…ah... Hey Ya…ah…"

Then he stopped. Quick as a flash, he reached back over his shoulder and acted as though he was removing an arrow from his imaginary quiver, which he placed in his imaginary bow, drew the string tight, and aimed it at a road sign. Suddenly he released it. His body froze and he became silent, as though he was listening to hear something. "Chief Cearley heap damn good shot," he announced without blinking an eye.

The heckler in the back just rolled his eyes at Randal's phony bragging, and the others tried not to smile.

Randal continued his story. "One time a little Indian boy wanted to know how Indians got their names. So, he asks, 'Daddy, where do names come from?' His daddy, Chief Fickle Feather, just looked at him and didn't answer. Finally, he said, 'You see my son, many moons ago the Great Father gave us every thing we have. He gave us the trees, the rivers, the fish, the fields, and all the game that we kill and eat. We Indians are happy people to have all these things. We think they are all very beautiful. So, to honor the Great Father, when a child is born, we look around and the first of these things we see is what the newborn child is named. You know Running Bear? Well, when he was born, his father looked out from his tepee and saw a bear running. You also know the one we call Rising Fawn, don't you? When she was born, the first thing that her dad saw was a small deer getting up from the ground.

"Chief Fickle Feather looked at his son and asked, 'Why do you ask this strange question, Pissing Wolf?'"

Laughter roared from the back of the crate, but Randal kept a straight face and didn't respond.

"Hunt-Uh! He didn't name him that, did he? Do Indians really get their names that way?" I demanded.

"Honest Injun. That's how Indians get their names. I wouldn't shit you, Little Joe. You're my favorite turd." Randal nudged me in the ribs, as he laughed, and said, "No. I don't know how they get their names, Little Joe."

The normal chatter of the bets and raises of the poker game had started and I overheard one of the guys holler, "Damn! Would you look at that luck. That f****r just drew into a Dead Man's Hand."

Then I overheard a familiar and hushed conversation ending with, "Watch this!"Hey Little Joe, do you know what a womb is?"

Instantly, my face began to turn red and burn. A weird sensation, I thought: I'm anxious, my ass is freezing and numb, my face is burning, but I have a good window breeze! My angst was rising, but I was all too aware of the upcoming routine. The catchers knew I was easily embarrassed, and they would say things just to see my face glow cherry red. The questions were usually about sex or other taboo topics, but I took the bait and responded, "Yeah, I think it has something to do with a woman's body."

To which the response was surly and challenging, "No!... No!.... Hell, No! You're not even in the right ball park. Why do you always have women on your mind, kid?" Then he announced for all listeners to hear, "What it is, is the sound you hear when an elephant farts in a cave. You know........., Ha...WOOO...... UUMM! "

He said this as he simultaneously slammed the heel of his open palm sharply against the metal roof of the crate. The old panel truck swerved and slowly recovered. Coins on the poker table went sliding everywhere. The players, almost in unison, began cursing the culprit and Frank Ray, the driver, displayed his rage, as he yelled "Goddamn, man! Don't do..ooh that! I'm still shell-shocked from Korea. I think I shit my breeches. I thought one of these four old 'maypops' we're riding on had just died."

Some laughed while the players continued to curse at the jokester, as they tried to locate their lost coins. Tensions often ran high inside the crowded crate. Verbal bantering and boisterous talk was the norm, but tonight the tone from a few of the guys was belligerent. The focus of the contentious talk was Carlie Tankersley. They thought he bragged too much and didn't work hard enough.

My face had regained its normal color as we drove by The Miami, but we didn't stop in this time. We had started late and the chicken truck was already setting at the chicken house waiting on us up on Hanging Dog Creek.

The stories behind the name Hanging Dog are many and diverse. All include a Cherokee Indian boy and a dog and they range from a hero Indian boy rescuing a dog to simply a dead dog being found stuck and decomposed in a creek. Most locals sided with the latter.

We arrived at the chicken house a little before dark and someone from the back yelled, "All right, Little Joe, get your scrawny bony little ass out of this crate."

Frank, Randal, and I bailed out and made our way across the rutted farm yard toward a chicken truck with the name, "Widow Maker," painted across the top of the grille. As the catchers stooped and got out from the back of the crate, a loud ass slap sound could be heard, followed by a commotion inside the crate. Someone had slapped Carlie a little too hard and a fight had begun. Frank sprinted back, stuck his head into the crate and yelled, "All right you sons a bitches! You know the rules: You fight, you walk home. I don't give a flying f**k who started it. What's it gonna be? Work or fight and walk?"

Carlie and Bill Brown had already stopped their wrangling, but that didn't stop Bill, as he emerged from the crate, to be heard muttering, "I'll get you later, mother f****r."

Two small houses of Rhode Island Reds were to be loaded on one tandem trailer truck. There were about two thousand from each house and we got busy breaking down the truck, as Frank put up the work light outside. Clouds sailed by in the dark sky and they sometimes blocked the moon's light. Randal was a great truck worker, but tonight he would need the light.

No one, with maybe the exception of Charles Cheatham, was better at working the truck. Cheatham only worked broilers, though, but Randal could work broilers and hens. Watching either of these guys fill and neatly stack the chicken coops was a thing of beauty. Their acrobatics and precise coop placement was amazing, because coop spacing was critical to the live arrival of the chickens at the processing plant.

We hadn't been working long, and it was really dark, when some one asked Charlie, "Hey, where's your look-a-like transistor sister?"

Charley, half joking, responded, "He went to Bomb Tokyo but a bear must have eaten him. No, he really did go to the woods to see a dog about a man."

When he hadn't returned after more than ten minutes, Charlie made his way over to me where he mumbled, "Hey Pissing Wolf. Come and go with me. Let's see what my idiot brother is into now." We sneaked out the back door of the chicken house, hoping not to be seen or missed. Frank was busy talking to the truck driver, so he didn't see us leave.

It was now pitch dark outside and at first we could see only the outline of the mountains, the road that snaked around them, and the other chicken house a little ways up the rutted road. But, we could hear Bill Brown and Bill Carter's voices coming from over near the edge of some woods. We also could hear a muffled voice that at first we didn't recognize. Then Charlie exclaimed, "That's Carlie. Run!"

Like two blind bats out of hell, we flew in our black-laced Keds toward the sound in the dark. When we got closer, we saw what had been going on. There was Carlie, his head was covered with a chicken feed sack, and his hands had been tied behind his back. Carlie was trembling, as he was held by Bill Brown next to The Pit. The lid was open and it stunk to high heaven. Brown and Carter had tried to shove Carlie, head first, down into The Pit, but had failed in their attempt. Carter had now pricked Carlie's arm with his knife, and had told him if he didn't go willingly, he would cut his throat.

"Let him go, Bill! We need him and both you two fellers back inside. We are getting behind and we won't get done before daylight if we all don't get back to work." I issued the order in a scared, but what I hoped was a suitably convincing voice.

"Not until I teach this son of a bitch to keep his f*****g mouth shut." Bill's words left no doubt he was serious, and his voice was cold as ice.

"Bill, I think he knows now, so let him go. I don't want to be here when the sun comes up trying to chase down these red birds. Come on. Let him go. I gotta go to school tomorrow." By this time I was pleading and didn't care who knew it.

Charlie was never at a loss for words but this time he was silent. Instead, he motioned with his head, and then he whispered, "Look, Joe. Over there."

I turned to my right a little and saw Garland Brown walking up to the Pit from Bill's blind side. Garland gripped a knife in his hand, and now stood about ten feet beyond The Pit. Out of nowhere, he had shown up just seconds after we arrived. The next words we heard was Carter's order to Garland, "Mind your own damn business, Garland. This ain't nothing to you."

Garland's response was simple and assertive. "Turn the kid loose and let's get back to work."

At that moment the clouds cleared away from the moon. Carter could now see that Garland, too, was holding a knife – an eight-inch pig sticker, to be exact. Continuing to look at Garland's knife shine in the moonlight, Carter finally backed down. "Okay, but you or somebody better tell this stupid son of a bitch he needs to keep his f*****g mouth shut."

Garland said nothing, but nodded in agreement. Brown released Carlie, and they both started toward the back door of the chicken house. Carlie's knees immediately buckled and he slouched to the ground. Charlie and I ran to him, and Charlie removed the sack from his brother's head and began untying him, while I held my nose and dashed over and closed the lid on The Pit. Garland left without a word, as mysteriously as he had appeared. He followed the other two guys back to the chicken house, but at a distance. No one, I noticed, put away his knife as they walked out of our sight.

As soon as the sack was yanked off his head, Carlie vomited. With a crying moan, he desperately tried to speak, but his words came out in wailing spurts. "I …could… be dead. I …wasn't going… in that Pit." Violent spasms of puking attacked him again.

I handed Carlie my handkerchief to clear the vomit, and told him that we didn't know where he had gone. I repeated again that he hadn't told us where he was going.

Still he lamented, "I… could be… dead. I COULD BE DEAD!"

After Carlie cleaned himself up, we helped him back on his feet and we all walked back toward the chicken house. The moving clouds had once again rolled in and covered the moon.

"Carlie," I said, "Go lay down in the crate. We'll tell Frank you're sick and need a break."

Carlie nodded his agreement.

Frank was still up at the front of the truck talking to the driver as we approached. He agreed to the break, but said Carlie would have to do some work or his pay would be docked. "Tell him to bring that chicken hook out of the crate when he gets to feeling better and catch strays. That'll be a little easier on him."

Charlie and I entered the chicken house again, caught a full hand of chickens, walked them to the truck, and acted as though nothing had ever happened out at The Pit. Carlie joined us within a few minutes and began working like a slave.

"Why didn't you take a break, Carlie?"

"I'm okay. Ain't no son of a bitch going to screw me out of getting paid on Saturday. I got a date with a sweet little thang down at the show Saturday night."

Frank soon returned from talking with the truck driver. Looking about him at the growing number of filled crates, he yelled, "Where the hell is Randal?"

Jimmy Limpey, who was now working the truck, shouted back, "He got chicken shit in his eye and couldn't get it out, so he took off down to the owner's house to see if they had a wash cloth and could help him."

Frank nodded. "Okay, but you need to open up that air space a little more on your next rict, so these chickens can get more air."

With a flashlight in hand, he strode into the house to see how many chickens were still to be loaded. Upon his return, he started slinging the chains over the ricts in order to secure them to the truck bed. And before he threw each chain, he yelled, "Headache."

Someone from the other side always yelled back, "Let her rip, boss man."

This routine was done to alert any passer-by on the other side that the chain was coming over. It cut down significantly on head injuries. Next he buckled down the loaded ricts of coops filling the truck bed. Nodding with satisfaction, he told Limpey, "Looks like we just about got this house loaded."

Within minutes, we were finished and the catchers all gathered near the truck, at the front of the first chicken house, and began to smoke their cigarettes. One of the catchers asked, "Okay, boss-man, what's next?"

Before Frank could answer, someone in the shadows hollered, "Ass-kisser..."

Frank ignored the comment and announced, "Listen up. I want all you pecker heads except Limpey, Monk, and White Boy to follow me up to the next house. You three guys stay here and finish cleaning out this one. We're going ahead to get the other house ready."

"Where the hell is Bear Shit Cearley?" another catcher yelled.

"I don't know," was Frank's response. "But I'm gonna' find out just as soon as we get this second house set."

We all followed Frank up a rough and washed out road. The gravel crunched under our feet, as our dark march stretched out toward the distant chicken house. I stayed close behind Frank. He had a flashlight. Carlie was on my right and Charlie was on my left. I had just moved right toward Carlie, in order to avoid an up-coming ditch, when I felt the crashing pain. It felt like my right cheekbone was being brutally forced over my eye ball and nose. At the same time, a slimy chip-filled substance penetrated my eye. Everything went totally dark and I screamed, "I can't see. I can't see!"

Unaware and thinking I was faking some drama, Charlie popped off. "No shit, Sherlock, I can't see either. You think it might be night time in Hanging Dog?"

By this point I was bent over and in piercing pain, but I didn't touch my eye. Carlie and Charlie stopped, and the march stopped. The road was blocked by the three of us.

"What the hell's the hold up?" someone from the back of the pack yelled.

Frank, not hearing any gravel crunch on the road, turned with his flashlight and shined it on me. He hustled over to me and asked, "What's a matter, Little Joe?"

I didn't respond.

Frank shined the light in my face. "Goddamn, kid, this don't look good," he exclaimed. He screamed, "Does anyone have a rag or something?"

Someone, I was never sure who, quickly removed his T-shirt and handed it over to Frank who started to try to remove the broken egg shell from my eye. "I better not try this here in the dark." He grabbed my hand as he asked, "Can you walk with me back down there to the owner's house? We can see to clean it up down there."

I answered that I thought I could, and we started retracing our steps back down the road. Frank turned and told the rest of the guys, "You all go on up there and start getting that house ready to load."

A voice in the dark blurted out, "Hell, Frank, we can't even see that house without some kind of light."

"Come and get my light," Frank offered. Suddenly I sensed, rather than saw, that he and I were in the dark ourselves. Frank led me slowly, in the dark, toward the lights of the owner's house. When we got to the front yard, an old blue-tick hound bounced off the front porch and started barking. I thought, "Oh, God, now I'm going to be eaten by a dog that I can't even half see." But, a man came to the door and his harsh voice issued an order. "Hush up, Blue!" Then, in a kinder voice, "Can I help you fellers?"

Frank explained, "This boy has been hit in the eye with an egg. I wondered if you could help me to take a look at it?"

The man agreed and swung open the door. "Sure, come on in and we'll see what we can do for him."

The house light was hard to adjust to, but I could now see from my left eye. We walked up on the porch and then on into a big open room. The furniture was sparse with only a few straight back wooden chairs and they were located in the corners of the room. In the corner to my left sat a young woman breast feeding a baby. A few other grownups were seated in the other chairs.

Everyone was very somber, and I felt like I had walked into a funeral home or had interrupted a church prayer meeting. A large table with benches around it was located at the other end of the large room. A single light bulb suspended by a cord hung from the ceiling over the table, and to my right, I could see a sink with running water.

Still trying to adjust to the light, I saw someone familiar through my squinted left eye. Seated at the table was old Randal with his head bowed. Before I could be certain it was him, Frank blurted out, "Randal, what the h…?" Then looking around the room he stopped and said instead, "Randal, you need to go help the other guys. They're up at the second house getting ready to start loading. They need you to work the truck."

As I walked toward the light, I heard a rather large-bosomed lady seated at the table say in tearful words, "That poor man! It's terrible, the cross that God's placed on his soul to bear." Her big knockers bounced up and down as she burst out in tears. Then she reached for a bulging burlap bag that lay on the table, and handed it to Randal. "Here brother, we want you to have this, too. It ain't much, but it will give them precious little young' uns of yours something to eat."

I glanced at Frank who looked stunned, but said nothing. Randal took the sack, his head hanging low and making a visible effort not to look at either Frank or me, he said, "Thank you," and headed out the front door, just a little faster than I thought was necessary. The old blue-tick dog barked only once, but then as best I could see, he followed Randal.

"Come on over here to the light, young man," said the owner. I walked over and stood by the sink. He sized up the situation and said, "I tell you what. I'll get you a towel, you stand over the sink there, wet it and see if you can get those egg shells out of your eye. I'm afraid it'll cut your eye if I do it."

I did as he instructed. It cleaned out quickly, but my cheekbone burned and stung with pain. More importantly, however, I could now see a little out of it, and felt a giant sense of relief that I wasn't blind. It didn't matter to me right then that it would swell closed the next day and turn black and blue. I wouldn't see out of it for about two days.

Frank thanked the family for their help and we started back to the second chicken house. The truck was almost loaded when we got there. We all soon loaded up in the crate and left Hanging Dog. It was a suspiciously quiet trip back home, but we were back well before the sun came up.

In a week my bloodied eye and cheekbone returned to their normal color. The small pieces of egg shell steadily worked their way out, and my vision came back. I never did know who hit me in the eye with the egg. Frank had confronted the crew about the incident. He made threats to them all, while brandishing a chain buckle in his hand, but, no one ratted out the offender. It was Code of the Chicken Catcher; a ratter would get his ass kicked and at the least expected time. I think I was a victim of collateral fire and the intended target had been Carlie.

The knife stand-off at The Pit had never been mentioned by anyone.

In a few days we also learned what had been in the sack that Randal left with from that house at Hanging Dog. The contents also explained why that old blue-tick dog had followed him. Randal had told the owner's family that his wife had recently died of cancer and that he alone was left to raise his five little children. He had also confided that one of the children had a heart condition, but that he had no money to buy medicine. The crying, big-bosomed woman had given Randal a twenty-dollar bill to buy the kid the needed medicine.

That burlap sack had contained a sugar cured country ham and a bunch of ham hocks. That's why "Ole Blue" had followed Randal from the front porch that night. Only Randal didn't make it home with the gifts. It was a week later before he even came home, and his wife had called Brownie to plead that Randal not be fired from the Crawford Poetry Company for being absent from work. She said Randal returned home broke, stinking, and had a bad case of the shakes.

Randal's powers of deception and persuasion were unmatched when his demons came to call on him.

7. Where I Come From

"But they said to Joshua, We are your servants. Then Joshua said to them, who are ye? And from whence come ye?"

Book of Joshua, 9:8
The Holy Bible

The chicken catchers all lived in a tri-state area of Georgia, Tennessee and North Carolina, known as the Great Copper Basin (GCB). Only five lived in Georgia: T-Bone, alias Glenn Watkins, lived down on Mill Creek, near me; The Tankersley twins lived up the creek a ways; Monk Holden, alias Ronnie Holden, lived in Baker-town; Ham Cearley, alias Randal Cearley, lived within a quarter mile of the Crawford Poetry Company in what was known as Villa North Town. Most all the others lived in Tennessee, in little company-owned row houses that surrounded The Great Copper Basin's mine works.

The GCB, located about 60 miles east of Chattanooga, Tennessee, and some one hundred miles north of Atlanta, Georgia, was a mining community. The GCB gained both international notoriety and recognition: Notoriety for poisoning the land, water, air and vegatation with its mining processes. Recognition for hosting part of the 1996 Olympics – the Slalom Canoe/Kayak events. During the heyday of The Poetry Company, however, it wasn't nearly as glamorous.

The original settlers of the GCB were much like the Cherokee Indians in one respect: they loved the land and distrusted outsiders. They both were suspicious of those who pushed for social progress.

Nonetheless, over the years, many people took jobs spawned by the mining and the processing of minerals. Many small towns formed around these works. Those who couldn't get a job with "the Company" or who didn't live on near-by farms, often worked for "the other company" – Crawford Poetry Company.

One such small town was Stafford Town, and it was located on a red gutted hill within sight of a copper smelter. My great-grandfather, Sam Crawford, had once been the minister at the church of Stafford Town. In the 1950s the old church building, a cemetery resembling "Boot Hill" on "Gunsmoke," and fewer than a dozen dilapidated houses were all that remained of Stafford Town.

Garland Brown was one of the catchers and he lived straight across the road from the Stafford Town church house. Garland had a ruddy complexion, wore a sandy, orange-colored beard, and he looked more Irish than a leprechaun. Garland was a true agnostic, but no one knew why. He distrusted a lot of people, but he was especially dubious of preachers.

Garland told me of a dialog he once had with an itinerant street prophet. The man was an evangelistic preacher, and he was preaching "the Word" right out in back of Ralph & Grady's Beer Joint, out there in the open parking lot. Garland said the eagle had flown that day, so he had downed a beer or two and was just standing there, out there on the Ocoee River bank, waiting for the local cabbie, Audie Hemphill, to pick him up. He said when the preacher finished his sermon, he had walked over to Garland and asked him, "Brother, are you lost today?"

Garland told the preacher, "No, I'm not lost. I live way up yonder on that hill." He pointed up toward Stafford Town.

Garland said the preacher was undeterred. "No, Brother, what I meant was, are you a sinner"?

Garland said he told the preacher, "No preacher, I'm not a Sinner. I'm a Brown. I go by Garland Brown. I live up yonder." Again he pointed toward Stafford Town. "Across from that God Damn church house."

Garland said the preacher turned and looked in the direction where Garland had been pointing, and said, “I’m going to pray for both you and that church.”

Garland said the preacher asked no more questions, but walked away and approached another person who was carrying a sack full of groceries. He asked them the same question.

8. *Dog Gone Hogs*

"It is quite possible that a man who gives less is more generous, if his gift comes from smaller resources"
Aristotle: Nicomachean Ethics

Catching chickens was my night job, but by day I was a farm boy. Daily chores and hard work were routine on our farm down on Mill Creek. On this day the routine changed and we loaded hogs instead.

Finding a hog pen located in the mountain woods is easy. A blind man could find a hog lot because hogs always seemed to be grunting and quarreling amongst themselves. You can smell and hear them from a quarter mile away on a windless day. The hogs we would load today would be no different.

Loading hogs from a pen to a farm truck requires following a few rules. The first rule is to stop the hog from moving in the pen. This is usually accomplished by catching the hog by its ears. Catching the hog by its tail will also work, but is not recommended for securing large hogs, because their tail will sometimes break off. Another rule is to always catch and hog-tie the largest hog first. This rule has nothing to do with respect for the senior hog. Instead, it has everything to do with conservation of human energy. By the time you get to the last hog to be caught, roped and loaded, you have very little energy left for wrestling a large hog. So, it is wise to make the last hog the smallest hog. Another rule for escorting a roped hog to the truck is to always walk behind the hog. A swine can be led nowhere with a rope. You must always follow behind the hog and make him think that he is the one doing the leading. Nudging him from the side to keep him on the trail is good, but you should never walk ahead or try to pull him. These rules cycled through my mind while traveling to Snake Nation, a rural north Georgia mountain community named by the Cherokee Indians centuries ago.

Snake Nation was where the hogs to be loaded were located, and the trip into Snake Nation was long and dusty back in the late summer of the sixties. The gravel road leading there meandered through the wooded hills and hollers of Fannin County, Georgia like a snake's path. Few farm houses or other landmarks surfaced along this red clay embanked road, and branch crossings with and without their pine slab board bridges, were a welcomed sight to travelers on this endless mountain road that began in Blue Ridge, Georgia. A surprised muskrat or even an over-grown weed refreshed the traveler of Snake Nation Road. Traveling this mountain thoroughfare on this hot Monday, our afternoon mission was two-fold. My older brother Jim, my buddy T-Bone Watkins, and I would load a farmer's hogs onto the truck in which we were all riding. A farmer living in Snake Nation needed his hogs taken to market, and my dad was responsible for the safe delivery of the hogs to the Murphy Sale Barn on the following morning. One other passenger accompanying us on the trip was an old, mongrel dog. He lay, subdued, in the floorboard. Our second purpose was to "let him out" somewhere around Blue Ridge, so he could easily find himself a nice new city home.

He was my old dog and he liked me and he was my trusted friend. He would look at me with sad, soft, knowing eyes, and that always made me feel good about myself. He trusted me and, next to T-Bone, he was my next best friend. My dog was of no particular breed and had no pedigree, but his colors of black and tan made him look like a high dollar coon dog. But to tell the truth, he was just more dog than anything else. I had traded a Barlow knife and an old number eighteen horse collar for him at the Murphy Sale Barn. I loved that old dog almost more than I loved my older brother, Jim, but my assessment of the dog had been all wrong. It had been with great sadness that I learned he was one of the worst kinds of a dog you can have around a farm – he was a "suck-egg dog."

We learned of his character flaw and his affinity for eggs one morning when our breakfast was interrupted by loud chicken chatter coming from the hen house.

"Listen, boys, do you hear that?" Mom demanded as she dashed toward the kitchen screen door that was held firmly closed by a long tensioning spring. "Something's after my hens" she screamed, as she pushed that screen door out of her way. She was running in panic towards the hen house.

Jim and I jumped up from the kitchen table and sprinted behind her into the frightful noise. My trusted friend had dug a hole under the hen house walls. We saw white egg shells stuck to his black lips and bright yellow egg dripped on his feet. Innocently, my dog looked up at me, showing no remorse. He began to wag his tail and his eyes seem to ask, "What's all the fuss about?"

Mom, who rarely showed emotion, screamed with anger, "Joe! Joe, you got to get rid of that suck-egg dog! He's eaten up all my egg money and we need that cash. My poor hens are scared to death now, and they won't lay another egg for a week."

Dad heard the story from Mom at suppertime and my dog's fate was sealed.

We let him out of the truck just outside the Blue Ridge city limits, before we started our long trip into Snake Nation.

Almost as if he were feeling sad for me and my loss of the old dog, Dad explained, "You know what boys? You just can't break a dog from sucking eggs once they've had their first taste."

Dad's words were well intended, but they did little to console us boys. Silently, we all turned our heads and looked back up the dusty road where the dog stood, watching his every motion, wondering what would become of him. He looked bewildered, but somehow, not lost, and he was sniffing everything in sight. He ran over to a tall poke salad plant, cocked up one leg and marked it. This was our last memory of my suck-egg dog as the truck rounded the next bend and he was lost to us.

Trying not to show that I was about to cry I looked at the floor and asked, "Jim, what do you think will happen to him?"

Don't know. What do you think he'll do, T-Bone?"

T-Bone just raised his eyebrows and shrugged his shoulders, but said nothing.

Dad drove the truck faster than one would normally drive on a mountain road, and he shifted to a lower gear each time the old truck started up a steep slope. He drove fast because he knew we had to get there and get the hogs loaded before dark. Dad also knew that most people in Snake Nation still didn't have nor want any electric lights. After riding for what seemed like hours, we now were looking down from yet another mountain top when we saw the rusty tin roof of a farm house where Dad thought the hog's owner might live.

"This looks like the place. Do you think this is it, T-Bone?" Dad asked.

T-Bone confirmed our hopes with a nod. He was familiar with the place because he had "folks" who lived in the Snake Nation community. The old house was a typical, poor mountain farm house; a clapboard, shotgun style house with a small front porch. The front door to the house was stuck half opened by a warped porch plank. Two large flint rocks stacked on top of each other served as the front porch steps. A wide washboard extended horizontally above the front edge of the porch and a wash pan hung from a rusty bent nail half driven into one of the rough-cut support poles. Paint had never touched this weather-beaten house.

The front yard was neatly kept. The red clay yard was bare of grass, had been freshly swept, and a chicken's foot print was all that could be seen to mar the perfection. The prints formed a trail that led to a Snowball bush located at the yard's corner, and under the bush, half buried in the yard swept dust pile, was a big old Dominecker hen. She occasionally fluttered one wing, attempting to inconspicuously dust her feathers with the small storm she was creating, but otherwise, she was quiet as a mouse eating lard.

As we parked and got out of the truck a man dressed in faded overalls emerged from the front door of the house. At the same time that he crossed the porch, he leaned forward and spit tobacco juice out into the front yard. Between discharges of the brown juice, he hollered, "Why don't ye come in and sit a spell?"

Dad told him, "We're here to get your hogs to the sale barn for you, and we'd best get started." At the same time, he looked back and yelled to us boys, "Go look for a good dirt bank to back the truck up to, so we can start loading."

Dad, was the king of chat and so, as always, he struck up a long conversation with the man while we put together a plan for loading the hogs.

Together we walked, scouting out a place for a good loading bank. T-Bone slowly surveyed the place and reached back to remove a thin, red and black Prince Albert tobacco can that protruded from a worn hole in his hip pocket. He stopped walking, quickly rolled himself a smoke, lit it up and studied the landscape. Jim continued walking and finally located a good bank, while I stayed behind with T-Bone and picked the cockleburs from my oldest pair of blue jeans.

"How many hogs this man got to be loaded?" T-Bone asked.

"Dad didn't say."

"I hope there ain't no sows with piglets; old sow will get after you if'n she thinks you gonna' hurt her babies," T-Bone informed me.

Dad finally stopped talking to the man and pointed to a small opening in the woods and shouted to us. "Over there's the trail to the pen." We all hurried toward a sloped foot path that led into the woods where the hog pen was located.

Remembering, T-Bone stopped and said "We'd better get the ropes."

We all turned back, retrieved the hog-tying ropes from the truck, and started back down the trail toward the hog pen. Dad backed the truck against the loading bank.

Grunts and squeals could be heard coming from the woods as we descended the narrow, rocky trail that led to the hog pen. The strong odor of the pen burned our noses and our eyes long before the pen came into sight. This was the easy part. Catching, roping and leading the hogs from their pen to our now distant truck was the hard part.

T-Bone, with his rolled smoke balanced between his lips, quickly leaped to the top of one side of the pine slab hog pen and squinted down gleefully at the nasty job that awaited us. T-Bone was older than Jim and me, but just how much older we never knew. Age was never discussed, and we all pretended to be older than we really were. T-Bone was tall, broad shouldered and lanky. He looked to be half Indian with high cheekbones, dark skin and dark brown eyes. At school, he had gained the reputation of a fighter, and he carried a razor sharp knife in his back pocket. No one had ever seen him use it on anyone. T-Bone had failed the sixth grade twice. He wasn't dumb, but just never could get to class regularly due to more important things he had to do. He was rough and tough, and he was our mission champion on this hot summer afternoon.

T-Bone quickly flipped his rolled smoke to the side and plunged, brogan boots and hands first, into the muck and mire that lay below at the surface of the filthy hog pen. The stinking mud splattered and some splashed into Jim's eye. The hogs froze in their tracks; silence set in. They looked scared and confused.

T-Bone grinned. "Come on in. The water's just right."

Jim nudged me and said, "Let's go, little brother."

With tie ropes in hand, Jim and I went up to the top of the pig pen wall. Seconds later, and much to the consternation of the hogs, we all chased after a jumbo sow hog in the muddy pen. We quickly loaded her and then we came back for the smaller hogs.

After the last hog was loaded, we three boys returned to make sure we had left nothing behind. As we stood at the now empty and quiet pig pen, T-Bone removed his can from his back pocket, took out the white paper cigarette leaves, and rolled himself another smoke with his smelly hands. Tired, filthy and rank with the smell of hogs, we made our way back up to the farm house along the same trail which we had individually escorted each hog to our truck.

When we reached the house something looked different in the yard and, as usual, my dad was still standing where we'd left him, in the yard, talking. The man in faded overalls was listening intently. At the far corner of the yard, however, near where the chicken had been resting under the Snowball bush, crouched an old lady dressed in a flowery, hand-made sack dress. She was hunkered down on her heels and appeared to be looking down into the dirt. As we approached the house, she didn't look up or acknowledge in any way that we were in her presence.

Still not looking up she spoke, suddenly, in a loud and chilling voice, "Verily I say unto you boys, you have all got 'The Mark of the Beast' on you."

It took a while for our young minds to discern the actual meaning of her "divine prophetic message." It finally sank in, and everyone broke out into laughter. The toothless old lady only grinned. Then, with no expression on her face, she looked back down at the dirt.

While standing in the yard and continuing to enjoy the laughter I noticed something coming from a long distance up the road.

It had begun to cloud over and it was starting to get dark. Through the haze, I could see a distant moving object and it looked like a small mule. As it continued coming toward the house, I couldn't believe what my tired, burning eyes were seeing. It was my suck-egg dog! He was traveling at a fast trot, but erupted into a full gallop when he saw me standing in the front yard. He ran straight to me with his tail wagging in a blur, and he started jumping up and down. Finally he stood on his hind legs with his paws on my soiled white T-shirt. He acted as though I had been lost, and he had found me and rescued me.

The laughter and talking both subsided, as the evening clouds darkened the front yard where we all stood. Everybody watched my dog. Even the old woman, still hunkered down on the other side of the yard, raised her head long enough to study the dog's behavior. The silence continued and it had now become deafening. T-Bone took a drag from his rolled smoke as he turned and looked back toward the hog pen. Dad's eyes sensed the tension building from the awkwardness of the moment, but he was uncertain what he should reveal about my dog. He said to the man, "That's a fine looking hunting dog you got there."

The old man firmly hooked his thumbs in the front buckles of his overalls, looked straight at my dad, and responded in a voice tinged with sternness. "I han't got nary dog and I don't reck'alec ever seeing him around here before."

As the man continued to study my suck-egg dog, he wiped the tobacco juice from the corner of his mouth and finally mumbled, "That dog shore is taken a liken to your boy, Joe, there. Why don't ye take him home waif ye for him?"

Showing some disingenuous consideration, Dad agreed to the man's suggestion. In short order, we loaded my dog into the front of the truck and headed out of Snake Nation as fast as we had entered it. The truck was fully loaded with hogs, smelly boys and one familiar old dog.

A slow mountain rain soon began to fall on the dark dusty road that led out of Snake Nation. My suck-egg dog once again rested his head on the floorboard of the truck. He was home.

9. Kidney Culture Conflict

"A man's gotta do what a man's gotta do."
Rocky Balboa: From "Rocky IV"

Bob Kisselberg was a farmer with land up on the Notley River and, like many farmers of the area, he was a proud individual. Now Bob was as mild mannered as he was tall, and he had a rugged build. His mountain man appearance was complete with a long beard. On his beautiful river bottom farm he kept laying chickens, hogs, and cattle, and he raised the food for his animals on that farm. Bob wore boots that really impressed me and some day, I hoped to own a pair just like them. The boots were shiny black, the same color of Bob's beard, and each boot had a chrome ring over each ankle where looped leather straps converged. A wide cross strap with a buckle connected to the rings. They were called engineer's boots, and Bob looked to be the king of the hill and master of his destiny in those boots.

Bob did have one little quirk. Not often, but on occasion, he would go to town to get himself a few beers and, after a can or two, he became a different person. It was like someone stole his true personality and put a phony one back in its place. He wasn't violent when he was drinking, but his attitude and demeanor changed. This normally passive, laid-back mountain man became a hyperactive menace to society and the Greater Copper Basin. Nothing or nobody – from a mangy mutt to Senator Zell Miller was safe when ol' Bob came to town. Like a spooked runaway plow mule in a knee-high corn field, he uprooted everything in his path.

However, Bob's farming enterprise had been one of good fortune and clever insight. His lush green farm was located remote, up-stream and up-wind from the killing toxins put off by the mining works of the GCB. Diversifying the animals grown on his farm, and crop rotation had been practiced on Bob's farm long before it was cool to do so. Bob Kisselberg was one farmer who not only managed to hold on to his lands and his income, but he prospered. Yet even that wasn't enough to insulate him and protect him from a tight spot that would prove to be an ultimate embarrassment to all his neighbors. Yet it never seemed to bother him.

My brother, Brownie, tells the story. He said it was late one Saturday afternoon when he got a phone call from Elmer Taylor, of the McCaysville city police. Elmer told him that he had locked up a guy by the name of Kisselberg who had requested that Brownie come down to the police station to see him. Elmer also told Brownie that Kisselberg had been arrested for public drunkenness and indecent exposure. Elmer explained that a well dressed lady, a member of the First Baptist Church, had been walking across the new Ocoee River Bridge, and happened to look down the river, where she saw Kisselberg taking a whiz off the side of the Old River Bridge into the Ocoee River. Indignant, she had marched straight to the police station and reported the incident.

Brownie said he went on down and Elmer let him in the cell to visit with Kisselberg alone, where the two briefly shared small talk. Finally Brownie asked him, "Bob, were you drunk today when they brought you in to jail?"

Brownie said Kisselberg became furious, tugged frantically at his beard, and then he snarled in protest, "Hell, no! I had only had a couple of beers at Ralph & Grady's. You know I got weak kidneys, and when I heard that river water running below me, I just had to take a piss."

Brownie said Kisselberg then took both hands and lifted a pant leg to the top of one of those engineers' boots he wore. "I didn't know what Kisselberg had in that boot, but I was dumbfounded when he pulled out a half pint of corn liquor. Bob shook it and held it up to the sunlight streaming through the cell window to check its bead. Then he handed it to me and said," 'But I guess

I'm drunk now. Have you a snort of this, Brownie. It's good stuff. I made it myself.'"

Brownie declined Bob's offer and told him that the preacher and his family were waiting to have supper with him back at home. Brownie bailed Bob out of jail, drove him back to his home near Culberson, North Carolina, and later helped him get his legal matters resolved. He said afterwards Bob didn't seem to remember any of the things he had done that caused him to land in jail.

10. Grand Ole Opry – Here We Come

"See the pretty snowflakes
Falling from the sky;
On the wall and housetops
Soft and thick they lie."
Anonymous: From "Falling Snow"

Catching chickens was not a fair weather job. It wasn't just seasonal work either, and harsh weather seldom stopped our labors. "Come hell or high water the feather necks must ride" was how one catcher phrased our mantra. But sometimes you had choices and today I had a decision to make.

Did I want to catch broilers with my brother Pete's crew or did I want to load heavy hens with Frank Ray's crew? The latter had an extra perk: I would get to go to The Grand Ole Opry. I was a bit torn by my decision.

Pete needed all the help he could get and that's why he asked me to work with his guys. His crew almost exclusively caught broilers and tonight he had thirty thousand of them to be loaded onto six trucks by sun-up. He was short on help, because two catchers were in the Benton Slammer and he didn't have time to go bail them out.

Frank's crew had only a partial truck load to catch – about one thousand heavy hens up around Culberson, North Carolina. After these were loaded, the small single axle tractor-trailer would return to Crawford Poetry Company's shed where cooped and waiting chickens there would complete the load. I chose to go to the Grand Ole Opry. I hated giving Pete a thumb's down, but traveling to Nashville to see the Grand Ole Opry was the deal of my lifetime. The chance to go there would never come around again, I thought.

The terms of the Opry deal were these: I would be a helper for the truck driver, Herbert Watkins. The helper's job included gassing up the truck, keeping it swept out, keeping the chains that secured the coops to the truck tightened, and making certain that no chickens escaped from their mobile prison. In exchange, I would get a Grand Ole Opry ticket, one hamburger, and one Coca-Cola. A handshake sealed the deal. I filled the tank of the chicken truck while Herbert finished tinkering under the hood. As we left the Crawford Poetry Company for Culberson, snow flakes darted and collided with the large windshield of the older cab-over Mack truck. We were North Carolina-bound… Bob Kisselberg's place to be exact.

We hadn't even made it to Mineral Bluff yet, when Herbert pulled the truck to the side of the road and stopped. He opened the truck door and began to dismount. "I kin make this thang run faster than e-is. I'm gonna take that there governor loose." He raised the hood and with a flashlight held between his teeth, he made an adjustment and climbed back into the truck. Rubbing his hands together with the attitude of a most satisfied man, he said, "Never would make it to Nashville before the Opry starts if I left that damn speed governor on there. It's colder than a witch's teat out there, Little Joe." Then he shocked me by placing one of his cold hands on the back of my neck and making a goofy sounding laugh as he watched my reaction.

Herbert was the oldest brother of T-Bone Watkins and he had just finished a two-year hitch in the army, where he had received training in jeep repair. He had no other education and hadn't even graduated middle school. Herbert was only in his late twenties, but already his hairline was receding, and he wore long black hair combed straight back. It always seemed to hang down in his stubbled face, however. Herbert's one distinguishing feature was his feet; they were extremely small for a man who stood over six feet tall. Benny, the middle boy of the Watkins family, would kid him about his feet and tell him, "Herbert, you ain't got feet. What you got is a pair of hooves like ole Maude." Maude was the Watkins' plow mule.

Herbert's small feet didn't slow him from getting around and, when he moved, he always seemed to walk off balance and looked as though he was falling forward.

His lack of education also didn't slow him from studying things of interest to him, but his studies weren't from a book. His favorite read was "Popular Mechanics" and his interests were limited to things mechanical, mostly because Herbert was a poor reader. He loved to analyze things to see how they worked, and he compensated for his poor reading skills with his analytical skills. Herbert worked hard to impress others, a determined man who had a burning curiosity for all things mechanical. He seemed to always be on a mission to learn.

We got to Kisselberg's place in less than thirty minutes, and it was the fastest ride I had ever taken in a chicken truck. Waiting for us were Frank Ray and a skeleton crew. Some of Frank's crew had chosen to work for Pete this night, because they could make more money. Still, we loaded that truck in less than an hour and we were soon back to the Poetry Company's shed. Frank and his crew followed us in the crate and they quickly loaded the full coops on the truck from the shed.

Now we were ready to begin our trip to the Grand Ole Opry, only we got no further than Gene Pack's service station, about two miles down the road, when Herbert said, "Reach behind your seat there and hand me them there bottles of alcohol. I'll get this thang running faster."

I did as he asked, and he took the alcohol, got out of the truck, and poured the contents of those bottles into the fuel tank. Then we pulled out, started down Harper Town Hill, and sped toward our destination, the Grand Ole Opry in Nashville, Tennessee.

Because I felt responsible for any escaping chickens, I kept a constant watch on the coops through the rear view mirror on my side of the truck. As we reached the peak of Boyd Gap and started down the mountain, I saw something I couldn't believe: A long blaze of fire was billowing upward from the end of the truck's vertical tailpipe. The chickens located closest to the flame were getting singed and covered with black smoke. Becoming more than alarmed, I instantly told Herbert about the flame.

Oddly, he seemed to be somewhat flattered by my concern, and he just laughed at me. "Watch e-is." Then he down-shifted the ten-speed Mack as we plunged from the mountain top. An even longer flame now burst forth from the tailpipe, and a brilliant blue and orange blaze could be seen piercing the black mountain skyline. The chickens shrieked with shock. They squawked and scrambled away from the mysterious blazing plume that jetted by their temporary sanctuary. I expected the chicken coops and the chickens' feathers to ignite and flame up at any moment, but they didn't. Herbert just threw his head back and laughed like a mad man as he returned the gear shifter to its original faster speed. Brushing his long, thin hair back with one hand, his face beaming with exhilaration and pride, he then asked gleefully, "What's you afeard of, Little Joe?"

We quickly finished our descent of the mountain and as we started into the curvy River Road stretch, the terrain began to level out. Herbert up-shifted the truck to still a faster gear. Reaching toward the center of the large windshield with his right hand, he pretended to be patting the head of a dog. It was in fact a dog; a chrome plated miniature one. He was patting the Mack Truck's emblem, a shining, silver bulldog located just below the truck's large glass windshield. Driving with his left hand and patting the chrome bulldog with his right hand, he said in a reassuring voice, "Good Doggie, Good Doggie."

When we got to Cleveland the snow was coming down heavy, and about three inches of fresh white snow covered the ground, but the roads were clear. We passed Payne Gap and I thought Herbert might stop yet one more time to make an adjustment on the truck, but he didn't.

Payne Gap was to Pete's crew what The Miami was to Frank Ray's crew; a place where chicken catchers' money was welcomed day or night. About a quarter-mile on down the road from Payne Gap was a large, white clay bank and below it, was a dirt race track. If a race was going on you could hear and see signs of the race track from Payne Gap. Red clay dust boiled and engines roared from the track. Some of the catchers liked to go there on Friday nights.

The next large city on our trip to the Opry was Chattanooga, but a few small wide spots in the road also lay between Cleveland and Chattanooga. One was Ooltewah, where traffic was detoured around Ooltewah because of road work. A new interstate highway was under construction. Interstate 75 it would be, and there was hardly any traffic out tonight to have to detour.

"I'm going to sleep a little," I said as I removed my Keds and climbed up in the center of the sleeping berth behind the seats. Instead of lying down, I sat up in the middle of the sleeper and looked out the big windshield at the peaceful sight of the snow that fell and blanketed the pasture land.

As the Ooltewah detour approached, I saw Herbert reach for a throttle knob he had located under the dash. Looking down at the knob he mumbled to himself, "I wonder what e-is does?" In a split second the truck entered a spiraling detour ramp with a slope that was sharper than either of us expected. The chicken truck was traveling too fast, and at first Herbert's side of the truck seemed to rise up higher than the passenger side. Next thing I heard was

Herbert's attempt to calm at least one of us. "Hold on, Little Joe, This thang's gonna …."

Those were also the last words heard. The next sound was that of metal dragging on concrete and the crash of glass breaking. I don't remember bracing for the crash. All I remember is becoming conscious of the fact that I was running as fast as my bare feet would carry me in the snow. I stopped running, turned around, and looked back at the crashed chicken truck as it now lay piled up in the middle of a snow covered field. I had run at least a quarter mile from the wreck. How did I get out of that crushed truck? Over the years I've asked myself that question many times, and have replayed the crash in my mind attempting to find the answer. Still, that answer eludes me, even now.

Then I thought, "Where is Herbert? Oh God, don't let him be under that pile of wreckage!"

Like a flash, I raced back toward the crash, but then I saw Herbert stumbling from the far side of the badly mangled truck. He was holding his head and was screaming. "Gee…it! Run! This thangs gonna' catch fire."

I slowed my run and Herbert limped from the wreckage. We ran toward each other, met, and stood together, each of us quiet, almost as though we were incapable of speech. Around us was the equally-silent, snow covered field there in Ooltewah, Tennessee. I remember looking at the now liberated chickens and thinking, "Why don't they run?" Their coops were busted open, they didn't appear to be injured, and they were free to roam. But they didn't run away. They just stood there and looked like a bunch of red tulips with their red-headed combs held high above the snow.

Soon we saw flashing red lights reflecting in the snow and my first instinct told me to run, but I didn't. A state trooper approached us, while back at the highway, rubber-neckers stopped to "help." A fire truck with more lights flashing and reflecting in the snow appeared next. Finally, an ambulance loaded us and took Herbert and me to Erlanger Hospital in Chattanooga.

I had several cuts, contusions and embedded glass in my body, but luckily nothing was life threatening. A foreign intern who spoke very bad English wanted to sew up all my bleeding cuts, but I convinced him otherwise. I told him that the cuts were only chicken scratches and that I got little cuts like these all the time. I assured him they would heal just fine without stitches. The truth was, I was scared of needles. Still am. He persisted, so I agreed to let him sew over a hole in the calf of my right leg, and it left a bad scar.

Herbert's cuts were less numerous than mine, but he had suffered a concussion. He apparently was unconscious for a brief period following the wreck. They wanted to keep us for observation and so, we stayed there at Erlanger and picked glass out of our bodies for a few days. We read comic books and played poker. One day Herbert "Mr. Tinker" was bored and wanted to know how his bed worked. He was curious as to how a bed could raise and lower, and be made to fold up, simply by pushing a button. I was dozing, but I saw him kneel beside his bed in his open back hospital gown and a nurse from the hall had also seen him when she cracked open our door and peeked in. Like a hummingbird, she flitted away because she thought Herbert was kneeling to pray. But he wasn't and it was an understandable mistake, since Herbert often mumbled to himself. He made a gallant effort to entirely disassemble his bed, but failed for a lack of tools. He was inquisitive and, if not for the pain, this man would remove his own eye from its socket and study its workings.

We played poker every day until I ended up broke. Even lost the hamburger and Coke I had been promised as part of the Grand Ole Opry deal. Then we played for match sticks. I won big time, and I left the Erlanger Hospital with seven boxes of matches and nineteen stitches in my leg. Herbert left enlightened on the inner workings of a hospital bed, and with an urgency to get back to driving and working on trucks.

11. *Stupid Is as Stupid Does*

"Life is Pain"
Buddha: The Meaning of Life

Herbert didn't get to drive a chicken truck again for the Poetry Company. His truck driving licenses were suspended because of the pile up, and the Tennessee State Patrol reported that he had been driving too fast for the road conditions. So, the Poetry Company made him a foreman and he started up another chicken catching crew. Not getting to drive again didn't bother him, because he really didn't enjoy driving a truck. What he truthfully liked was working on that truck. Also, Herbert was married now, and driving a truck meant being on the road and not seeing his new wife.

Herbert asked me to join his new crew and I did. We usually would catch heavy hens over around Gainesville or Buford or Hiwassee, Georgia. I liked working with him because the crate trips were shorter, more money could be made, and the River Road wouldn't be on the route. The Poetry Company rarely loaded heavy hens in Tennessee. Also, I liked working in his crew because his younger brother, T-Bone, was a crew member, and he was closer to my age than most chicken catchers.

After getting married, Herbert rented one-half of a duplex two-story house that was across the road from the McCaysville First Baptist Church. The house was said to be "half in Georgia and half in Tennessee" because its builder had made a mistake in locating the house: he had placed part of the house on the state line.

In that house one Saturday, I learned some painful lessons that would stay with me for life. It was near Christmas time and I had made over twenty dollars working in Herbert's crew that week. On payday I was proud and excited to get those earnings, because twenty dollars was the most money I had ever made in a week for catching chickens. It also meant I would have plenty of money to spend at Christmas. But that Saturday turned out to be both a horrible and unforgettable day for many reasons.

I lost every cent I had earned that week in a poker game that lasted less than an hour. The day was also unforgettable because of the weather. The entire Copper Basin area was a giant chunk of ice. The day was the coldest winter day I had ever spent in my life in the Copper Basin area: The temperature was about twenty below, snow drifts were piled over waist-high, and the wind howled all day. No feather necks would roll today. All roads were closed and engines refused to crank and run. All traffic had stopped moving.

But the harsh weather didn't stop me from getting to the poker game at Herbert's house across from the First Baptist Church. We had talked about the game all week, and I had looked forward with great anticipation. So I endured a frigid two-mile walk to get to the game. The return trek seemed endless and cold, and I was filled with regret and angst and misery. On that walk home, I felt like the dumbest, coldest, most alone stooge in the world, because Christmas was coming and I had lost my Christmas money. Money that I had planned to use to buy presents for Mom and others. Walking home, I slid and fell on the frozen ice several times, but I reasoned that the stinging pain was just punishment for a dumb loser like me.

"How could I have been so stupid?" I thought out loud and I wanted to cry, but I didn't. It was too cold to cry and the tears would have just frozen to my eye lashes. That would have shown everyone what a real loser I was. Worse than being a loser, I thought, was people knowing and laughing at how stupid I had been. I was cold and broke, and had nothing to show for a week's worth of hard work. It surely was one of the worst days of my life.

But slowly, I got over the guilt from losing the money, and my lost self-esteem returned. I shook off the shame like it was frozen snow stuck to my boots, and I went back to work determined to put it behind me. I even tried harder in school as my mom and my teachers had always advised me to do. "Who knows?" I told myself. "Maybe they're right. This thing called education might be a good thing." So I resolved to not play anymore poker for a while, and to never call a double-or-nothing bet in a strange game called, Black Jack. Instead, I would watch and learn from others.

That frigid Saturday had nothing to do with it, but the desire to go to the Grand Ole' Opry had left me. I was into "Rock and Roll" now and Country Music and their performers' clothes seemed so corny and out of touch with the times. Their stories' lyrics, I understood and could easily relate to them, but I thought it strange and depressing that anyone would set them to music and sing about them.

Later in life my curiosity of country music returned, and my wife and I agreed to take our children to Opry Land. For the first time, she and our children all got to see a Grand Ole Opry performance. We even took a nostalgic tour of "Do-Wa-Diddy-City." I told them all about the chicken truck wreck at Ooltewah, and they thought it funny. Deep down, I knew they couldn't relate to that story, so I also told them about me losing all my Christmas money one icy day, and I think they got that one. The warm day in Nashville was one of the best days of my life: Feeling no pain and seeing my family having fun at Opry Land!

12. King Tommy

"All the world's a stage, and all the men and women merely players: they have their exits and their entrance;"
William Shakespeare

We didn't load up the crate and go catch chickens every night. When things were slow, I sometimes worked during the day at Crawford Poetry Company's base, where the company's office and a garage were housed in a small two-story building. Downstairs there was a shop set up for maintaining the chicken trucks, and my brother Jack managed that part of the business. The second level, up a stoop, was the business office. A large gravel lot surrounded the building where chicken trucks were parked. A large, high bay shed also stood in the center of the gravel lot. Chickens were weighed and damaged chicken coops were repaired and dispatched from this shed. My brother Brownie and his family lived in an old house that adjoined the gravel lot.

Just down the hill, and across a red clay dirt road from the back of the old house, lived a young boy named Tommy German. A narrow trail forked off the dirt road and led from his house up to Brownie's back yard. The business office was just a few steps on up to the top of the hill. A flurry of activity and a cloud of gravel dust were ever present at the base for the Crawford Poetry Company. Business men, chicken catchers, truck mechanics and chicken truck drivers were in and out of the place at all times.

Brownie's house had always been called the "Old Johnson House" – Johnson being the name of the previous owner of the old house. Doris, Brownie's wife and bookkeeper for the business, was intent on restoring the old house. Renovation was constant at the old structure, and no one was ever at home there during the day. No one ever noticed Brownie and Doris's old house.

Now Tommy German was different. He was a boy of ten and was full of odd. He looked and he dressed, he even walked differently than most other kids in our small neighborhood of Villa North Town, Georgia. A wistful, perplexed, and yet determined expression could always be found on Tommy's face. Strangers thought him mentally challenged, but that wasn't the case. His burr cut hair revealed his oddly shaped head, but also revealed that he kept himself clean and neat. Tommy wore his pants a little higher on his hips than most boys would find comfortable, his walking gait was fast, and he often would randomly jump into a skip. Tommy marched to a drum beat that only Tommy had ever heard.

I caught a glimpse of Tommy one morning while working atop a ten-high stacked chicken coop truck. From this lofty lookout, Tommy was seen hastily ascending the wet and slippery trail leading up to the old house. He walked and then he skipped across the back yard, and around to the front door, where he inconspicuously entered the old, unoccupied house. Unknown to all was the fact that Tommy religiously made this sojourn every morning, come rain or come shine, and what Tommy did during his daily morning stay inside the old house, was known only by Tommy. That would change on this bright, sunny summer morning.

Tommy quietly closed the big door behind him, turned around and knocked three times on the closed door. Then he followed a hallway that led to the den area, where a door on the right led into a large room. This large room was now under renovation, but no one was working on it this day. Tommy was careful to step only on the black tiles of the black and white checkerboard patterned floor tiles that covered the floor.

From the den he wasted no time opening the door leading to the large room to the right. A relieved look crossed his face when he saw the now familiar things in the room. Objects he had seen many times before on his many early morning visits. The large over-sized room contained all the equipment and appliances you would expect to find in a laundry room – a clothes washer, a clothes dryer, shelves for storage, and a shower stall. One other object was centered along a distant wall. This object, a large, white ceramic toilet, was the point of arrival for Tommy's morning quest.

The space of this large, oversized room defined beauty and solitude for Tommy, but one thing was conspicuously missing and Tommy failed to notice. The room had no exterior walls. Gone also was the roof over the room, and the morning sun highlighted the bright, white ceramic toilet. The renovators had removed the walls and the ceiling the previous day. The oversized and now exposed room assumed the surreal appearance of an amphitheater stage. The various appliances were the props for the upcoming play, and the now-prominent ceramic toilet sparkled as the rising morning sun beamed down. It looked like the throne of a King.

Tommy didn't see the missing walls and the absent roof. He had never really seen them when they had been there. They were unimportant; not in the script, and certainly not a part of his morning agenda.

Tommy quickly approached the ceramic throne and he dropped his much too high short pants and sat down. Propping his hands under his now quieted face, he looked down at the floor never noticing that the renovators had also removed the linoleum floor covering. Tommy didn't see the now exposed floor boards. A mockingbird sitting on a nearby empty clothes line began to imitate a cat bird. Looking up and seeing the bird, but still not missing the walls, Tommy mimicked the bird that became spooked and fluttered away. Evidently feeling sufficiently relieved, Tommy sat up straight and began singing, "I'm Popeye the Sailorman, I'm Popeye the Sailorman, I'm strong to the finish 'cause I eat me Spinach, I'm Popeye the Sailorman…, Beep, Beep."

He stopped singing and he placed both hands high above his head and he began a war game wherein each hand was a bomber plane trying to gun down the other one. Several sorties later, the battle was over and Tommy resumed his rendition of the "Popeye the Sailorman" tune.

He was totally oblivious to the traffic, and to the people entering and leaving Brownie's office. Some people could hear Tommy but they didn't see him on stage. Others saw him, and they would grin as they walked by him, quietly and quickly, with their heads down protecting his privacy. But no one stopped to catch the next act at the morning amphitheater.

Mission accomplished, Tommy finally pulled up his short pants extra high and buckled his army surplus green belt. Then he exited the stage in exact reverse order, retracing the path he had traveled to his morning destination. As he left, he was overheard whistling his strange rendition of the Popeye the Sailorman tune, as he skipped gaily down the slopping back yard trail, back to his own home.

13. I B Driv'N

"Can a man take fire in his bosom, And his clothes not be burned?"

Book of Proverbs, 6:27
The Holy Bible

The Crawford Farm was down on Mill Creek and was bordered by the Old Epworth Road on the Georgia side. The farm was also bordered by "Fighting Town Creek," and it was covered with mostly wooded hills and a few cleared pastures and corn fields. The family farm had once been a dairy farm where my older siblings, including my brothers, Brownie and Jack, who now ran the Crawford Poetry Company, had all worked hard to develop the Crawford Dairy. In today's jargon, it would be called a "24/7 operations" because work on the dairy was never ending. Some of my siblings said they were glad when lightning struck the dairy's main milk barn and it burned to the ground. I remember the farm differently.

For a pre-teen boy, the farm was a place of adventure and enlightenment. The dairy was gone when I was there, however the 24/7 chores and responsibilities remained. But time was always made for some fun, too. The farmland raised corn and hay, and farm animals that were usually delivered and sold for cash at a stock sale barn. My dad, Carl Crawford, was a horse trader and earned money by purchasing and reselling farm animals. He sold milk cows and plow mules to "country folks" who had farming in their blood, and they lived on the outskirts of the Great Copper Basin. Because driving a farm truck was essential, my dad took me to get my drivers license when I was only thirteen. Getting a license was no problem back then; all you needed was to pass a driving test and have a Social Security number, which was also no problem to get. Passing the driver's test was a snap. I'd been driving on the farm and even on the back roads by the age of eleven. Driving was freedom and I loved it!

On this night we were to catch chickens down below the mountains at a place in Bradley County, located between Old Fort and Cleveland, Tennessee. The chickens were to be delivered to the Campbell's Soup Company in Cleveland, Ohio. The truck driver was Carl Ray, and I was getting to go with him as his helper. Monk Holden usually traveled with Carl, but not on this trip. I was really excited!

The feather nest was totally different. It wasn't a chicken house at all. Instead, it was an egg factory where each B-42 was individually caged in what one catcher called "a little jail cell." In the front of the jail cell was a conveyor belt that carried chicken feed and water. At the back of the jail cell another conveyor belt that carried away the eggs the jailed chickens had laid. Underneath the jail cell was yet another conveyor that carried away the poop of the incarcerated.

The B-42s in the jail cells looked and acted a little different from the B-42s we normally caught in a regular feather nest. These chickens looked like my best teacher ever, my eighth grade teacher, Mr. Chastain. Both he and the chickens wore a thick set of spectacles and, in a half-assed way, this made them look intelligent. But they weren't really glasses. Instead, the apparent set of glasses was just a barrier that was strapped to their beaks. The guy who worked at the feather nest said the glasses were blinders that kept the chicken from ever seeing the eggs she produced. Without these glasses, he said, the chicken would peck and destroy and even try to eat its own eggs.

Unlike other B-42s, these confined chickens were aggressive and ill natured. In a New York second they would peck you and flog you if you tried to remove them from their little jail cells. But if you dropped one, she didn't instantly take flight and try to escape like regular B-42s, either. These jailed birds acted like they didn't even know how to fly.

We finished loading the truck just as it turned dark. B-42s were never caught during day light hours from a regular feather nest. Because these were caged up and essentially already caught, loading during the evening hours was fast and easy. Carl and I chained and buckled down the truck and headed out. The road from the egg factory was level and well-graveled, and was nothing like the roads to most mountain farm feather nests. Within a few minutes Carl and I were on the paved highway-"Highway 4-11," headed toward Knoxville, Tennessee.

Carl was a great truck driver and he made driving look like a game instead of a job. He also didn't mind telling you some of the secrets of his driving skills. Gears were changed by Carl with the same finesse and ease as that of a magician pulling doves from a top hat. He was a master of his trade.

We hadn't been on the highway more than an hour, and we were rolling along on this long, straight stretch of road, when this big black Cadillac sock-top with the top down, pulled up alongside the chicken truck. From the passenger side of this truck, I couldn't see the driver that well, but I did manage to see that two women were in the front seat. I also saw the woman on the passenger side of the Cadillac hold up ten fingers. Carl nodded his head, as if in agreement to something, then the two women sped away, as I watched those red Cadillac tail lights in the big fins fade out of sight.

We pulled over at the next truck stop. Carl slid the big rig up to the gas pump, shut off the engine, then he looked at me, and said, "Okay, Little Joe, here's what I need you to do. I want you to fill this thing up with gas, check the oil, and put some oil in if it needs it. Then I want you to go inside and get the receipt. Don't worry about paying for it. Crawford Poetry Company has an account here. Be sure you get the receipt." He handed me a dollar bill and said, "Get yourself a Coke and a candy bar if you want one. Now, here's what else I need you to do. I want you to drive the truck on up to Dixie Lee Truck Stop. I'm going to meet you there in about two hours."

"Shocked by the last part of his request, I protested. "Did you say for me to drive this truck? I've never driven a tractor trailer truck like this one on a highway before."

Before I could say another word Carl attempted to assure me. "Look, I know you can drive. I see you parking and pulling up chicken trucks back at the shed all the time. You have drivers license now, and I see you drive your dad's farm truck all the time. I know you can do it. All you got to do is change the gears and keep the truck on your side of the white line. You'll do just fine."

Carl jumped down out of the truck and joined the two ladies in the Cadillac, that was parked over near the highway.

I did everything exactly as he had asked, including getting myself a Coke and a Butterfinger candy bar. Then, with just a little less confidence in myself than Carl had, I fired up the truck and struck out up the highway north, headed for Dixie Lee Truck Stop, about forty miles on up the road. There was only one small town to go through, where I would have to stop, shift the gears and bring the truck up to speed. As my chicken catcher buddies would describe my emotional condition, "I was as nervous as a whore in church." My big worry was being stopped by a Tennessee State Patrol and maybe being taken to jail. I did have a license, but it was only a learner's permit. Georgia allowed you to drive with them for one year before you got a real set of driver's license. Fortunately, I had no problems and pulled the chicken truck into Dixie Lee Truck Stop in about an hour's time, where I waited for Carl to complete his other mission.

When I stopped the truck at Dixie Lee, I got out and walked over to a bench where an old man was sitting in front of their garage. He looked me over and sized me up. Then he said, "Son, anybody else in that chicken truck you just pulled up in?"

I told him, "No! But I'm waiting for someone."

His comeback almost unnerved me. "Son, you look mighty young to be driving a big truck like that. How old are you?"

I lied and told him I was sixteen. I also wondered if I was about to get busted somehow.

At about the same time, out of the corner of my eye, I saw a Collie dog walking away from the parking lot. The dog had a slick coat and was pretty, and looked like Lassie on TV. It walked over to some really green fescue grass that was growing over past the edge of the parking lot, and that dog started eating grass. The dog ate almost as if it didn't much like grass, but it ate it anyhow. To change the subject, I quickly asked "Do you see that? That dog is eating grass like a cow or something."

My interrogator looked over at the dog and said, in a manner that indicated he was totally unconcerned, "Yeah, there's sum'pin it ain't get'n. That's what makes em act that way."

We talked a little more and we mostly talked about dogs and goats and fishing and farm animals. He didn't ask and I didn't tell him that we were going to take the B-42s to Cleveland, Ohio, to the Campbell's Soup Company. In just a few minutes Carl pulled up with the Pavement Princesses in the Cadillac. He got out of that big Cadillac and sauntered over to where I sat. "You ready to go, Little Joe?"

"As ready as I'll ever be."

He scratched his chin. "Let me get me a cup of coffee and off we'll go to Oh…Hi…Oh."

As he went inside, I stood outside by the old man and thought to myself: I don't understand anything that's happened so far tonight. I especially don't understand why Carl, my hero, took off with those two whores. The dog eating grass made more sense to me than what Carl had done. I'd seen Carl's wife, and she was pretty and smart, and had been a great basketball player in school. She had one other arrow in her quiver that my chicken catchers had often noted: "She was built like a brick shit house."

This was just the beginning of a voyage into the bizarre. Driving and freedom, and seeing new things for the first time were exciting. But, for a farm boy from down on Mill Creek, the meanings of all these things were not clear.

14. A Bowl of Diversity Soup

"Two things teach man about his whole nature: instinct and experience."

Blaise Pascal

Balancing a paper cup of coffee in one hand, Carl readjusted the driver's seat and both mirrors, and we left the Dixie Lee Truck Stop. Handing the hot coffee cup to me, he said, "Hold this a minute for me, "Little Joe."

Once he had brought the chicken truck up to speed, smoothly working his way through the gears like the true professional he was, he retrieved his coffee and we were Cleveland bound. Our next big town was Knoxville, but I had been there before. The Crawford Poetry Company delivered broilers to Harrison's Processing Plant just off Magnolia Avenue, and I had traveled there once with another driver.

Knoxville had a peculiar odor. It smelled like hot metal and yet, it smelled like something sweet too. I was told that the smell came from Alcoa, an aluminum plant in nearby Maryville. Just beyond Maryville was the airport, and it was near the highway we traveled. At night, its runway was a stunning sight. Ground level rows of red lights and blue lights and green lights seemed to extend into infinity from the highway. The road that we traveled through Knoxville was narrow and difficult to maneuver in a chicken truck, but Carl made that obstacle course look like child's play.

As we left Knoxville I began to wonder, but didn't ask, why Monk Holden wasn't traveling with Carl on this trip. I had known Monk for several years and really liked him. He was more than worthy of his apish nickname – Monk was absolute proof of Darwin's theory on the evolution of man. The Scopes Monkey Trial up at Dayton would have been short-lived had Monk been the defense's "Exhibit A." Summary Judgment would have been ruled in favor of the defense, even before the jury's *voir dire*, had Monk been in the courtroom for that trial. Monk's appearance was capable of casting doubt in the mind and heart of the most fervent "Creationist." This young man was Ug-leeee!

Monk's physical appearance was what the outcome might have been if a female gorilla had been crossbred with a cave man. His facial features were that of a monkey. His forehead was short, and his eyes sunk deep into his skull. His teeth were covered by long thin lips and they protruded from his flat face and his big-holed nostrils like a small round melon.

His arms were long and hung almost to his knees. Monk's arms would probably not have appeared so long if he had stood erect, which he didn't. He was slightly humpbacked and, when he walked, he looked as though he was falling forward.

Monk's skin gave you an eerie feeling, if you looked at it long enough. His complexion was almost albino, except for the rusty red freckles that covered most of his exposed body, and strangely enough, matched the color of his thin sparse hair. His skin's tone looked like that of an amoeba, a shrimp, or some other critter that had crawled from its aquatic environment and was now forced to live on the parched earth.

He had one endearing quality, though. Monk was super-funnee. His style of humor was not self-deprecating or slapstick, as you might think for one with his appearance and background. His dad worked for the McCaysville Sanitation Department. Though Monk had quit school after the seventh grade, he was smart and thoughtful. He was also witty, and it was he who often diffused tense moments among the catchers, thanks to that trait.

Monk lived with us down on the farm on Mill Creek for about two years. My dad gave him room and board and ten dollars a week to do work on the farm. He cleaned out stables, repaired fences and anything else that needed to be done around the farm.

He was a good steady worker. At first, he lived in our small, one-room smokehouse. Monk jokingly referred to that room as "the Rat Run Inn." Salted and sugar-cured pork had once been stored there, and Monk usually smelled like that smokehouse. Later, when it got cold outside, Monk moved into our house and he and I shared a bedroom. Well before the rooster crowed each morning, Monk's gigantic "Big Ben" alarm clock religiously awakened us and together in the dark, we'd go feed the farm animals before we had breakfast. After breakfast we would slop the hogs. We took all meals together, and he was always great fun at the evening supper table.

Once, my mom, Nona, became bored, and tired, just plain burned out with the constant drudge and daily monotony of farm life. In her frustration, one night at the supper table, she announced, "I've got to get out of here, I can't take this anymore. I'm going somewhere!"

Monk, who was always quick on the uptake, picked up on her frustration and asked her, "Oh, is that so? Where are you planning on going, Nona?"

Mom, who was now becoming even more aware of her dull and hopeless situation, snapped back in a choppy, fast-paced voice. "I don't know, I just don't know. I may go to…I'm going to… 'The Country.'"

Without looking up from his plate of beans and taters and cornbread, Monk said in a slow and serious, deliberate voice, "Now, Nona, you be extra careful when you go, and when you cross the creek over there at Allie Watkins house you be real careful. That fast moving subway over there might run you over."

We all laughed out loud except my mom. I laughed too, but as I laughed, I wondered – what's a subway? She jumped up and left the supper table in tears.

My dad just shook his head as he said, with amazement registering on his face, "Going to The Country? The Country! The

Country, my ass. It don't get more country than living down here on Mill Creek."

Monk was afraid he'd hurt Mom's feelings and offered to go apologize, but Dad told him not to worry about it. I felt sorry that my mom was so down, but Monk's deadpan comment was just too funny to not laugh.

Monk got a job that paid better with the Crawford Poetry Company and moved away from the farm. He went to live with Carl and his family and he traveled with Carl on the chicken truck. But for some reason, he didn't go with Carl on this trip to Cleveland. Instead, I was Carl's helper this time.

After we left Knoxville, I decided to get some shut-eye. The air was still warm and moist, so I just left the window down. I felt safe in the chicken truck, and felt even safer knowing that Carl Ray was at the wheel. I laid my head against the truck's door and went fast to sleep with a soft breeze blowing on me.

"Are you a strong swimmer, Little Joe?" were the next words I heard.

I mumbled, "What? Where are we?," as I wiped rain from my face and head. We were in a terrible thunder storm and the rain was coming down in buckets.

"We're almost to the Kentucky line," said Carl in an unusually loud voice, "but I don't know if we'll make it there if this keeps up! I've never seen a storm quite like this one."

About the same time the road and the whole sky became as light as day. I saw a possum on an urgent mission quickly trotting across the road, its tail held straight out. A large strike of lightning had just flashed and instantly, a thundering boom vibrated the inside of the truck. I was scared! I was also getting a chill and was trying to get myself awake. I was, as my chicken catcher buddies would say "shaking like a dog shitting a peach seed and didn't know whether to shit or go blind."

I looked back at the B-42s on the truck to quickly see how they liked the storm. Rain had never touched them in their entire lives back at the egg factory, and they were settled into a pile at the back of their coops, attempting to dodge the strange matter that pelted them. I wondered how the possum and other animals would keep from being drowned.

We were in the mountains on a curvy two lane road. Enclosed by the towering peaks, it seemed like we were traveling in a large, twisted dark tube that was being filled with water. The rain streamed down the mountains onto the highway and it carried leaves and sticks with it, along with anything else that got in its way. A dirty, white-capping river gushed across the road at places where the pavement sloped. I wondered, "Is there any place on earth where it isn't raining? When will we ever get out of this twisted dark world of water?" Then, up ahead, I saw a bright white road sign with large black words reading "Do Not Pass." How dumb and totally irrelevant, I thought! Silently I prayed, "Please God! Get us to the other side of this dark and wet wilderness."

In no time at all, the rain began to let up and the road even straightened out a little. You could see some gaps between the mountains. "Pass with Care" the next road sign read. A line of cars that had followed us through the flooded mountains immediately started doing just that. The next sign read "Welcome to Kentucky, The Blue Grass State." I was now able to relax a little. I turned to see how the B-42s were now riding – absent of their newly discovered phenomenon, the rain. They had separated themselves from their piles and were moving about and life was more familiar to them, and they seemed more at ease. But then, they didn't know that within forty-eight hours they would be chicken noodle soup, either.

Carl pushed the old "Widow Maker" up the road as fast as its ten gears and the Kentucky speed limit would allow. Our next stop was a weigh station just across the Ohio line, where the ICC would check out the rig, the chickens, and Carl's Log Book.

We crossed the Ohio River at sun-up and the golden glow shimmering and glinting off the water was a beautiful sight to see. Large barges loaded with coal and other goods were being pushed up-river by little towboats. Why, I don't know, but all I could think of was an image from a story my teacher had read to the class. The book was Uncle Tom's Cabin and the image was that of a black woman escaping to freedom by jumping across frozen blocks of ice that floated down this same river we were now crossing. I still have trouble identifying with someone with that much guts.

As we started on to the Ohio River Bridge Carl announced, "We're getting ready to go into a strange country, "Little Joe." You can't understand much that these people say on the other side of this river. They're Yankees, you know."

I wasn't totally ignorant about people from Ohio. I had once visited some of my cousins who lived up in Akron. Funny thing, I thought, about what Carl had said. My cousins all thought I was the one who talked strange. When I visited them, they invited all their friends over just to listen to me talk. They all laughed at me, and even my relatives laughed at me. I thought they were rude, and that they didn't know much about anything important. None of them could plow a mule or milk a cow. They didn't know how to whittle out a flip stock or even know how to shoot a flip. Heck, they didn't even know how to catch spring lizards for fish bait. And all they knew about chickens was that Mom brings chicken home from the grocery store. They had a lot to learn, I thought.

Before we got to the weigh station Carl saw a sign on the road that told him the scales were closed. We wouldn't have to stop, and Carl was glad, because his log book wasn't up to date and, by law, he had driven longer than he should have without stopping to sleep.

Cincinnati was interesting. The houses all seemed to be narrow, but very tall with most of them being three or four stories. They had small windows and steep roofs. The houses were all built close together and were painted loud colors. Back home, down on Mill Creek, the closest house to us was T-Bone's house, over a quarter of a mile away. Cincinnati also had many different scents, thanks to the many factories scattered around the city. We drove by an Avon perfume factory – it was the king of scents. I also saw the place where Pepsodent and Colgate toothpaste were made. Altogether, it took about two hours to get through Cincinnati. Once we were on the other side, Carl called for a stop. "I got to have some sleep and something to eat. We'll stop up here at the Buckeye Truck Stop and get some breakfast. I'll give you a little money and you can play the Pinball while I catch forty winks."

As we pulled into the Buckeye, I was awestruck. It was BIG! The Miami, Payne Gap and the Dixie Lee all combined would cover less space than this truck stop. Trucks with tags from all over the USA were here. Tractor trailers like I had never seen before were everywhere, and some of the tractors pulled two trailers that had been hitched together.

The place was sprawling. It was packed with activity. The Fannin County Fair couldn't match this place for frenetic activity and dazzling lights. The place had spaces allocated for special services. A place was there for putting air in your tires. A place was there for washing your truck. A place was there to just park and sleep. And, like all truck stops, there was a place to fuel up. That was our first stop.

Carl said, "You go ahead and fill it up and check the oil. I'm going inside. I'll clear it with the attendant and pay the bill. Park it over there in the Sleeper Lot when you've finished, and come on to the diner. I'll have some breakfast waiting for you. I'm going to get a shower."

I did as he asked and then joined Carl. The diner was also enormous, with swivel stools that lined a counter bar and they looked to be three miles long. There were at least ten waitresses, all dressed alike, working behind the counter. Tiny little juke boxes were spaced out about twenty feet apart on the counter top. Large tables were also there for groups of truckers that traveled together in a convoy. I found Carl near the front, with an empty stool and the promised breakfast waiting.

Handing me a dollar bill, and looking at his watch, Carl said, "Here's you some money for the Pinball. I'm going to go sleep a few hours in the truck. If I don't come back by two o'clock, you come and wake me up. We still got a long ride on up to Cleveland. Oh, don't forget to check those chains and catch any chickens that get loose."

As Carl turned to go, I saw the most peculiar human being I had ever seen. He walked along beside Carl. He was a black man and he was also a truck driver like Carl. The strange part was, he was Carl's twin, only he was black. He was a little on the chubby side, but he had a friendly smile and had told some of the waitresses, "I'll see 'you skies' on the flip-side." His teeth were pearly white and his height and build, and even his cocky walk, was like Carl's. He wore a professional truck driver's hat – one that had a shiny black brim and a crossing chain, similar to the hat worn by a five star general or high ranking police officers.

One thing sharply distinguished this guy from Carl, however. It also distinguished him from anyone else I had ever seen in my whole life up to this point: This black guy had blue eyes! They were bright sky blue with long eye lashes. I couldn't help but stare at him.

Carl later explained to me what I had seen. He said that the guy wasn't all black. Carl said he was what was called a "blue gum n*****" and had some white in him. He said the man probably had a white mom. Blacks sometimes marry whites up here in the North, he told me. I was from Mill Creek, and had seen only a few black people before, but I thought to myself, "Yankees sure are strange people. Don't matter if they are black or white."

I did everything Carl told me to do and I finished my breakfast. Then I played Pinball. Since this place had about a dozen different machines, I won one free game. After the money ran out about the same time my interest did, I mostly just walked around and looked at all the different types of trucks. It was like livestock competition day at the County Fair.

Carl didn't come back so I went out to the Sleeper Lot and woke him up at two o'clock. You had to be very careful when you roused Carl from a sound sleep. You also had to be careful not to sneak up on him. If you walked up to him on his blind side, he would holler, jump and sometimes strike at you. So, I just banged on the truck door a couple of times and he sat up. As he climbed down from the sleeper, he said, "Check those chains while I go take a whiz." He soon returned with a cup of coffee and we loaded up and headed for Cleveland.

I don't know what time we rolled into the Campbell's Soup Company in Cleveland. When I woke up it was dark outside and was drizzling rain. I just went back to sleep. Carl was sound asleep on his side of the truck.

I was awakened just before daylight by a loud screeching, metal-on-metal sound, but when I looked out, I couldn't see anything but tall, white concrete retainer walls. They surrounded us on all sides and it was like the truck was setting in the bottom of a large, concrete bowl. The rain had stopped, so I climbed out to see where the noise was coming from.

Still unable to see anything, I decided to climb on top of the chicken truck. The B42s were not alarmed by me climbing up and over their new home and, at the top of the coops, I stood up and could now see over the concrete bowl. From there I could see what looked like railroad tracks, but it seemed to have an extra rail. In less than a minute I heard something coming down the tracks so I got low, out of sight on top of the coops, and peeked out over the top of the concrete bowl.

What I saw was one more thing that I had never seen before. It appeared to be a train but this train was different. The train cars weren't hauling pulp wood, like the ones do over at the Pine Ridge, Tennessee wood lot. The train car wasn't an acid tank either, like the ones back at the Copper Company. This car was hauling strange looking people and you could see them as clear as day because the train car was made of mostly glass.

The people it hauled looked like "Hop Sing," the cook on Bonanza on TV. But these people didn't have a braided pig tail like that little Chinese man. Their skin was pale and they looked spooky, and I didn't know quite what to do. So, I opened the top of a few chicken coops, gathered up a pile of eggs, and began throwing them at these strange looking people. The eggs made a great yellow splatter on the glass side of the train. Seeing they were under assault, the strange people sat up straight, looked all around, but they could never see exactly where the eggs were coming from. The strange-looking train was traveling too fast. After I ran out of eggs, I climbed down from the top of the chicken truck, woke up Carl, and told him what I had seen. I also confessed to what I had been doing. He just laughed out loud and said that what I saw was the Cleveland Subway. The people, he explained, were Chinese, just like Hop Sing was, and that a lot of them worked for the Campbell Soup Company.

About an hour later the Cleveland police drove up in a car that looked like a Holstein cow, and they looked all around. They talked to a Campbell's guy for a while and finally drove away. I stayed slumped down in the chicken truck so they couldn't see me, but I got out when they were gone.

We were the first truck to be unloaded at Campbell's that morning. I put the tarp over the front rict of empty coops of the chicken truck and strapped them down with the rubber tarp straps. This would make the truck run faster, and by mid-morning, we were on our way home. In about twenty hours I was eating breakfast that Mom had made for me, and it was great to be back home down on Mill Creek.

Carl got some terrible news when he got home. His wife had left him. She had filed for divorce so she could marry Monk Holden. She and Monk had planned the whole thing. That's why Monk hadn't made the trip to Cleveland. Carl didn't have a clue and was blindsided by the whole affair. He was devastated for a long time, but he was never violent toward Monk.

As is the case for most divorces, the unintended victims were the children. My parents had divorced too, so I shared the pain and confusion they felt. But Carl dearly loved his children and, in order to spend time with them, he quit the Crawford Poetry Company to drive a short haul tanker for the Tennessee Copper Company.

Just as Carl's divorce created permanent pain and, to some degree, hardship, his situation was not unlike that of the Great Copper Basin. The mining and smelting of the copper ore took its toll on the ecology and Carl's absence at home had taken its toll on his wife and children. Acid rain had denuded plant life in the GCB and steady winds had eroded the soil's essential nutrients. Red and barren hills defined by huge gullies now flourished where fertile farms and trout streams had once existed. What once was, would never be again. Carl's once beautiful family and the GCB of years past were only a fading memory. Once again, while I understood what had happened, there were so many things that were still unclear.

15. With Liberty and Justice for All

"Mankind is at its best when it is most free. This will be clear if we grasp the principle of liberty. We must recall that the basic principle is freedom of choice, which saying many have on their lips but few in their minds."

Dante: From "Monarchy"

Work on the farm was not hard, but it was like the air we breathed, in that it was always there. Every monotonous day, it was there. The farm animals depended on us for their food, their water, and their shelter, since they were fenced in or were confined in stables and couldn't access their own needs. They depended on us, and we didn't let them go without. A rule on our farm was: "The animals eat and drink before we do." The first and last chore of the day was to see that they had been fed and watered. In the spring and summer, some of these chores went away when the animals were "turned out to pasture," where they were free to meet their own needs. Our farm had good pastures and Mill Creek ran through those grazing spreads, so the animals were fat and happy during these months.

But our work didn't stop because, during these same months, we would plant, cultivate, and harvest corn and hay that would be the animals' food supply for the coming fall and winter months. We spent long days in the corn and hay fields, and they were usually spent at the end of a hoe handle or following behind a plow mule. This was the work pattern on the farm, and the fact that you caught chickens didn't change that obligation. But mom would do our chores for us if my older brother, Jim, or I caught chickens at night. After the night I'll never forget, this pattern would forever be changed.

On this night, we rode the white horse with glassy blue eyes down the grass-laden hillside to the meadow. He carried Jim and me on down toward Mill Creek. It was autumn and a cool breeze chilled us, but it was the springtime in our green lives. No saddle, no blanket, no flashlighta leather bridle and two May-morn boys was the steed's total burden. In the pasture we looked for a black cow with a white face that our dad would take to the Cleveland sale barn the next day. He needed cash money – a last minute decision was our dad's modus operandi of recent.

At the supper table a man-to-man discussion had just taken place. The topic was about Jim leaving the farm. He would go to live with Brownie's family, and would sleep in a small bedroom in the office of the Crawford Poetry Company. Jim would earn money by working on the weekends helping my brother Jack service the chicken trucks. The details had all been worked out, and Mom and Jim had talked it over in private, before Dad came home that evening.

Jim was the boss on the farm. He worked faithfully and could do just about anything that needed doing around the place. But now he had some dreams of his own he wanted to pursue. Like my older brother, Gene, Jim hankered to play football. Gene was a former captain of the football team and football scholarship player at Tennessee Tech, and Jim could see himself following in Gene's footsteps. His leaving home was to be expected though, because most all the Crawford boys left home early. Some left even before they graduated from high school. Pete, Bob, and Gene had all left to live with my sister, Johnnie Lee, and her husband, Myles. Jim wanted Dad's permission to follow his dream, but Dad's acceptance was what he really wanted. Dad's permission was nice but not necessary. That's how Jim really felt about it. Dad knew personally a thing or two about a young man having his dreams blocked, because as a young man he had left home and joined the army. But he had been kicked out and sent home when the army learned he was only sixteen years old.

I didn't want Jim to leave and, as the white horse slowly marched us down the hillside, I told him so. I felt that he was abandoning me. I wanted him to be happy, but if Jim left, all the farm work would become my responsibility. Was I ready for this big change, I asked myself?

But it was more than that. I wouldn't have anyone to talk to anymore. If Jim moved away, who would answer my questions? Those important boy questions... like who would win in a fight between Roy Rogers and Gene Autry?... Is country music really here to stay?... Or why do girls look funny when they run?... And why did we have so many brothers and so few sisters in our family? Also, I thought that he would no longer be my brother if he moved away.

At school, he would be just another older kid and, maybe, he would act like that little adopted girl at school, the one that always ignored me when people whispered she was my bastard half-sister. She heard the whispers too. We looked at each other, but we never talked to each other. In my class at school, I was small. If Jim left, he might no longer protect me from the older bullies at school. Maybe. But mostly, I feared that Mom and Dad would be disappointed in me when they saw that I was not able to do all the things that Jim could do on the farm. I felt small and alone, and thought Jim's decision was selfish.

It was ironic that I didn't want him to go. Truth is, we constantly fussed, argued and fought. Usually it was about nothing. We sharpened each others metal, and our adversarial behavior was normal and to be expected between brothers, some would argue. Well, maybe they were right, but I felt we were total opposites – like salt and pepper or oil and water. I also felt that Jim, with his blond hair and great All-American Boy looks, was Mom's fair-haired child. I was just the runt baby of the family, and I had always felt this way, until years later, after I overheard my mom say, "I've got nine children and nary two of them are anything alike. But that boy, Joe, he's my independent one. He thinks he's got to do it all by himself."

Jim and I had totally different personalities and now Jim was becoming different in another way. He was becoming a man while I was still a kid. Jim wanted to make his own decisions, and he did. He moved away. I wanted to slug him for leaving me, but I also wanted to hug him. Confused and sad I also wanted to cry about my situation, but I didn't. No crying was allowed. Cry is a word not found in the chicken catcher's code.

After Jim left, my work load on the farm went up and I didn't get to make much money catching chickens anymore, because Dad really needed my help. But I made some money raising my own veal calves. Driving the farm truck and hauling farm animals was great fun, too, but I missed a lot of school to help my dad. He was getting older and his strength and verve were fading.

On Thursdays in the late fall, we usually took a load of plow mules and horses to the sale barn in Chattanooga. It was down in St. Elmo, at the foot of Lookout Mountain. Big company buyers came from up north to buy these "long in the tooth" horses and mules. They paid about six cents a pound for them, then hauled the over-the-hill steeds back up north in large covered livestock trucks. I once asked, "Dad, what do they do with all those old jar-head animals they buy?" He told me that some were slaughtered to make dog food, but that most were taken to the "glue factory" and were used to make glue. I wished he had lied to me instead.

One Thursday at the St. Elmo sale barn, I was just kicking around the grounds by myself and Dad was up front, talking to someone before the auctioneering began. I was too far away to hear their conversation, but I was close enough that Dad could see me. I had a sack of roasted peanuts and was shelling and eating them, while I looked at all the different horses and mules, when an old man sidled up to me. He looked like he hadn't shaved or taken a bath in about three hundred years. He also gave off an air of total exhaustion. Looking down at the sawdust on the ground, he asked, "Can I have one of your peanuts son?"

I looked at him closely at first. "Sure. Hold out you hand and I'll give you some."

He didn't say anything else. And he stuck the peanuts I'd just given him into his pocket, but then he pulled one back out and examined it. He kept looking at that peanut like it was a cat-eyed marble, or something like it. Finally he shelled it and ate it and, without another word, just shuffled off.

When we started home from the sale barn later that afternoon my dad asked me, "What were you and Mr. Simmonds talking about?"

"Who?"

Dad explained that Cy Simmonds was the name of the man to whom I had given the peanuts. He said that up until about three months before, Mr. Simmonds had been a wealthy millionaire from over around Sugar Valley, Georgia. It seems that he and Brownie had shared in one joint venture called C&S Hatchery. He went on to say that Mr. Simmonds suffered some unexpected bad business luck, and had lost everything. He even lost his home and had suffered a nervous breakdown. A few years later my brother, Brownie, suffered a similar fate.

I had thought Mr. Simmonds to be just another horse trader with some serious trading on his mind. But then I remembered having caught broilers with my brother Pete's crew over at one of his chicken houses in Sugar Valley.

In February I spent the entire month with Dad in Florida. Mom did all the chores on the farm while we were gone. Dad had been hired by a wealthy Yankee retiree, who now lived over near

Bell Hill Straight on the Tennessee side, to investigate the profitability of buying cattle in Florida, hauling them to Tennessee and selling them there. We went to cattle sale barns all over the state of Florida. We recorded the cost and price elements of every aspect of the Florida cattle business. But mostly, we had a great time traveling around in the Florida sunshine. We ate a lot of oranges and bologna sandwiches. Dad took his girl friend, my step-mother-to-be with us. I took my school books and tried to teach myself.

It was March when we returned and because spring planting was coming up, Dad said that I should not go back to school. He wanted me to stay home and work on the farm. His orders were confusing to me but I always tried hard to do what he told me to do. So, like some of my other siblings had done, I quit school to work on the farm.

At the end of the school year, at a time when no students were present, I returned to school to get my report card. Mr. Chastain, my eighth grade teacher, met with me and told me about my grade results. He was a great Georgia History teacher and reported that I had very good grades for the small amount of time that I had attended. He said I did especially well in Algebra. Then his tone changed, and he went on to explain that I had missed more than sixty days of school. Under Georgia School Law, and despite my good grades, I would be required to repeat the eighth grade. I had failed at school for the first time in my life, and I felt like a dumb country hick. How could I face everyone next year?

Mom moved away soon after Jim left home. She went to live with Brownie and she worked in the Crawford Poetry Company Hatchery. Mom and Dad had been divorced for several years, but had continued to live together for the sake of the children. My step-mom-to-be immediately moved in with us on the farm. She treated me good, but I didn't always get along with her kids. Her son "Buggs" was younger than me, and was a good kid. We got along well. He loved living on the farm and never complained about all the work. I taught him how to drive the farm truck. His older sister, Pat, on the other hand, was a pain. No one could get along with her, including her own mom. Eventually her mother sent her away to live with her older sister in Gastonia, North Carolina.

In the fall, when school started back, I found myself in the eighth grade for the second time. It was embarrassing at first, but I liked that I was now as big as the other kids in my classroom. In the summer of the same year, we had planted and cultivated about thirty acres of corn. It was now time to start gathering the crop, and Buggs and I pulled corn every evening after school until dark. On those few days, when it wasn't raining, he and I would get excused from school at mid-day, so we would have more time to work. Still, the barn wasn't filling up fast enough for Dad, so he told me and Buggs to stay out of school the entire following week. I didn't like my dad's demand, but I said nothing. I also read his demand as a sign that he had no plans to allow me to go to school anymore. Ever. So, like all my brothers before me, I made plans to leave the farm. I shared those plans with no one.

My sister Johnnie Lee had agreed instantly that I could come to live with her and Myles. No questions were asked. I will never forget my escape from the farm. Instead of going to church with the family on Sunday night, I told Dad that I needed to do some work in the barn. He agreed. Work trumped everything on the farm, even worship. While everyone went to preaching, I hauled my meager clothes and my other belongings over to Johnnie's house using the farm truck. I then returned the old farm truck, parked it, and I ran away from that dark holler down on Mill Creek. I ran toward a new future.

I felt shameful about leaving my dad, because he was my dad. I loved him dearly. The guilty feeling still haunts me. Abandoning your dad with a failing farm in the autumn years of his life is a chicken shit thing to do, but the circumstances left me with no clear excellent moral choice. Prayers didn't provided a clear one either, but they did promise me assurance in living with the consequences of my decision. My Decision. But the choices were all dappled-like that gray horse that pulled the corn gathering wagon. And what of this thing I had chosen – this thing they called "Education"? I knew of it, but only as a fish knows of water. What's it any good for and good compared to what? Uneducated chicken catchers and educated teachers alike had told me that I needed it. All said they wished they had more of it. So maybe, just maybe, I had made a good choice. The future would tell, I figured.

16. A Natural Lie

"...the law is a ass – a idiot. If that's the eye of the law, the law is a bachelor; and the worst I wish the law is that his eye may be opened by experience – by experience."

Charles Dickens: From Oliver Twist

When I moved in to live with my sister, Johnnie Lee, and her husband, Myles, I had to accept changes in my life habits. Becoming properly domesticated wasn't easy for me. The main change was in the area of hygiene. Johnnie Lee didn't just believe that "cleanliness was next to godliness." She thought that cleanliness WAS godliness, and all people living under her roof would act godly. She was the Queen of Clean. Down home, on the farm, I had taken baths, but usually only when I had been catching chickens, loading hogs or if we were going to church. With Mom gone from the farm, I was free to take a bath at my own choosing, and this amounted to taking a bath only when someone told me I stunk. Dad usually told me when the stench was bad enough, but now Johnnie Lee told me. She scrubbed on me for about a week straight when I first moved to her house to stay, and she would pull and bend my ears and inspect my scalp daily to check me for cleanliness. It made me feel like I was a dog being checked for ticks and fleas. It was embarrassing and aggravating, and I didn't like it. My liberty was lost, I thought, but in truth, the deal wasn't half bad: Stay clean and you get a place to sleep and something to eat every day! I slowly learned that cleanliness and Johnnie Lee's house rules were a fair compromise.

The house rules were simple. A bath shall be taken every day was Rule #1. It mattered not that you caught no chickens that evening, or that you weren't going to church, or you simply felt that you didn't need a bath. A fresh, daily change of clothes was Rule #2. My sister Johnnie Lee washed and ironed clothes constantly. When my brothers, Pete, Bob, and Gene had all stayed at her house, she did more laundry than either of the two laundries in the Copperhill area. Rule #3: Shoes were never to be worn in her house. Bare feet on her spotless waxed wooden floors were moderately acceptable, but she preferred that you wear white socks. If your shoes had any mud on them, they were to be removed and left outside the house. It was kind of like living in a Japanese home, but without having to wear a kimono.

At Johnnie's house, I was free to catch chickens as much I wanted, but only if it didn't interfere with my school studies. Myles usually kept me on track with my school work and, now, it was a breeze because I had been enrolled in a class with the not-so-smart kids. The school placed me there because I had failed the eighth grade and, school policy dictated that I be put there. But they had moved me to a less challenging program because of the failure, and it was pretty boring. It mattered not that I had been tested the previous year and found to have the highest IQ of all eighth graders. It was nothing personal against me; it was just policy.

Tonight, I would work with my brother Pete's crew. We were catching broilers near Ringgold, Georgia, at a feather nest the catchers called "the Elbow House." The feather nest was long but was not straight, and its shape was in fact like an elbow. The owner had built it that way to avoid having to grade out from the base of a hill. "The Elbow House" fit between a curved creek on one side and the bottom of that hill on its other side.

All the feather nests located around Ringgold were uniquely similar in one way: they were heated with steam pipes. Six to ten pipes spaced about four inches apart ran the full length of one side of the feather nest. The layer of pipes was mounted about two feet above the floor. The broilers could walk under the pipes to get warm; similar to how baby chicks would gather under the wings of an old free-ranging mama hen. This unique heating system caused big problems for the chicken catchers, but that problem wasn't the danger of being burned by the pipes. Instead, the problem was that the broilers would hide under the pipes, and you couldn't reach them, without first lying down in the chicken manure in front of the pipes. The pipes were often cursed and kicked, and about every catcher at one time or another was temped to frell hell out of those damn pipes.

Pete came by Johnnie's in the crate and picked me up first. The round-up circuit was "back-ass-'ards" this night, because some of the catchers living in Newtown had told Pete they wouldn't be home. They wanted to be picked up last. So, we went toward Isabella first to pick up the catchers that lived over that way. After leaving Newtown, we would start our trek to Ringgold.

Traveling to Ringgold was always a dreadful trip. The path taken was "The Old Copper Road" – the same route as the one built by the Cherokee natives over one hundred years before. The route that led from Copperhill was narrow, curvy, and had only two lanes. Sometimes the trip made you toss your cookies. Passing slow moving Cowboy traffic was near impossible on "The Old Copper Road," and horrible wrecks often occurred, due to failed passing attempts. This was especially true on the part called "The River Road." The Ringgold Road from Ooltewah to Ringgold wasn't much better. This road was not so curvy, but it had dips and several elevated, one-lane bridges that were like launching ramps. A ride on this road was like a carnival ride on a roller coaster. The crate's tires would sometimes leave the ground if Pete forgot to slow down at the appropriate time, as he approached one of these bridges. Not slowing down also upset the perpetual poker game ongoing in the back. If the crate traveled too fast across these bridges, the coins and the cards on the coop bottom card table would be launched into orbit. It didn't do anything for the attitudes of the chicken catchers, either. Wrecks were the norm on the Ringgold road, and head-on crashes usually occurred on, or near, these one-lane bridges, because approaching cars would race to be first.

This time we didn't even make it to Isabella before we had to stop. When we got to Five-Points, we were not able to turn toward Isabella at the intersection. Traffic flow from all directions was stalled out.

The "Goat-man" was parked there, at the intersection of highways 64 and 68 and the Burra Burra Mine entrance road. The place was a glorious circus! Tourists and local drunks and a strange cross-section of all God's humanity seemed to be gathered. A large herd of goats of all possible sizes, shapes, and colors were milling around a heavily laden wagon that was pulled by teams of goats. Some cars pulled over to have their inhabitants'pictures taken with the "Goat-man." Now this ol' boy looked ROUGH. He was old and scraggly, and needed a hair cut worse than Rip Van Winkle. He had a long, grey beard and he wore a striped train engineer's cap. His only clothes were a pair of overalls, that looked as though they had been thrown away about forty years before and hadn't been washed since.

Some people stayed in their parked cars and just rubber-necked out their rolled down windows. Some women and kids seated in the cars, and some of those who stood around the goat wagon, held their noses. Was it the scent of the goats or the nearby chemical plant's So2 gas in the air that caused them to hold their noses, I wondered. Maybe it was simply the odor from the "Goat-man" himself.

Pete somehow managed to pull off the pavement and get around the traffic snarl being caused by this one-time nomadic personality. We were once again on our way to Isabella, up toward Bell Hill straight where the guys with rods raced and sometimes wrecked. But when we got to Isabella we didn't stop. We took a left and kept going over toward Stansbury Gap.

"Where we going, Pete?" I asked.

"We're going over to get Jackie Cavender. You know Jackie don't you?"

"Yeah, I know him. Dad trades horses and cows with Harrison, his dad. He's the one who sold Dad "Big Red," that big Roan cow that had those little twin red calves. How long has Jackie been in your crew?"

"Not too long. Dad asked me to put him on a few months ago. Jackie's a good worker for a little guy" Pete explained.

"Little guy" was the under-statement of the millennium. Jackie was a midget and was hardly four feet tall. He was even shorter than me. But he was built stout – especially in his upper body. He used that upper body strength to sometimes maneuver his entire body, like getting into and out of a car. His biceps looked like those of a weight lifter, and they extended out from his thick chest, instead of hanging down. Strangely, he was well coordinated and could turn a backward flip without any effort. His legs were extremely short and were the most disproportionate part of his body. Jackie's pants never seemed to fit him right, and he appeared to waddle when he walked.

On our way to Jackie's house, Pete related a conversation that had taken place about a week earlier between old Ham-bone and Jackie. Pete grinned as he told how old Ham-bone had tried to give Jackie some legal advice. He said the conversation took place while the catchers were all standing around, waiting for the chicken truck to arrive at the feather nest. Pete said old Ham-bone began giving his legal advice to Jackie by saying, "Hey, Jackie, I think you ought to sue the City of Ducktown. You've got a perfect case. You could walk away a millionaire when you win your case, and you wouldn't have to catch chickens anymore."

"What the hell are you talking about Ham-bone?" Jackie had demanded.

"Well, Jackie, you see, the City of Ducktown has encroached upon your due property rights. You are legally entitled to be compensated for their wrongful actions. All you need to do is take your case to court."

Unconvinced, but still intrigued by the possibility of coming into a lot of money, Jackie now looked seriously at Ham-bone and asked, "What property rights are you talking about Ham-bone?"

Without any change in his tone of voice or change in his facial expression, Ham-bone explained, "Jackie, the encroached property is your ass. You see, the City of Ducktown built their sidewalks too close to your ass."

"Cearley… you can just kiss my little biddy ass," is all Jackie said, as he turned and waddled away in his poorly-fitted pants.

Those listening started to laugh and got a big bang out of Ham-bone's insult, but they caught themselves and stopped immediately. They realized the code had been broken; you don't demean people about things they can't change – like their size. Hambone knew he had violated the chicken catcher's code and would get his ass kicked later by the others, if he didn't make amends with Jackie. Later, Ham-bone apologized to Jackie and they became good friends.

I really liked Jackie. He was good natured, had a quick smile, and worked hard. The catchers tried to cut Jackie a lot of slack because of his size, but Jackie didn't want slack. He was proud and wanted to be treated no differently than anyone else. Pete usually had Jackie to catch the chickens and then have the other catchers to take the chickens from Jackie's hands and carry them on out to the truck. Pete told Jackie that this would make the chicken's get loaded faster, so Jackie was fine with his assigned responsibilities. Jackie's hands and fingers were like mine… despite all my stretching exercises, my finger were destined to remain short. We both had a problem sometimes with dropping the chickens. Jackie, because of his diminutive stature, however, was especially good at getting the broilers caught from under those aggravating pipes over there in the Ringgold feather nests.

When we left the pavement on our way to Jackie's house, we soon realized that a vehicle was ahead of us. We couldn't see it, but knew from the cloud of road dust that it was there. The road leading up and through Stansbury Gap was well graveled, and a few small farm houses bordered the road. You could travel fast on this road even though it had only a single lane. Room was available for two cars to pass, but only if one pulled over and stopped. Pete slowed the crate to let the dust settle, so we wouldn't have to inhale the choking mess.

Just before the road began its incline up the mountain, and at the beginning of a curve in the road, we saw a commotion. Smoke and steam could be seen in the distance boiling up through the thick, dusty haze. As we traveled closer we could see all too clearly the tragic sight ahead – one that I will never forget. Two cars had collided. It was obvious that there would be injuries.

The car we had followed, a 1949 black, four-door Ford, was involved. Its occupants were a family of five. The husband had been driving, his wife was on the passenger side and three children, all who appeared to be under the age of six, were riding in the back seat. With them in the back seat were the shards from several broken, half-gallon glass jugs of raw milk. Glass and milk were mixed with blood, and the children were wallowing around in the glassy, sticky liquid mixture. Two children screamed hysterically for their mom, but she didn't answer. The oldest child was silent and their mother, I realized, was unconscious and motionless in the front seat. Her white blouse was covered in blood from a gashing head injury, and her right eyelid appeared to be totally severed. I could see her eye and where once the lid had covered it. The stench of hot transmission fluid and radiator steam permeated the air. I wanted to throw up, but didn't.

Pete, who was trying to help the mother, brought me back to reality when he barked, "Quick, Joe! Take off your T-shirt and give it to me. I've got to get this bleeding stopped. You get in the crate and go back to that last house we passed. Have them to call an ambulance." He glanced at the five occupants of the car and seemed to make a snap-judgment. "Make that two ambulances. And tell them to hurry! Go on, now!"

I did exactly as Pete had asked. Luckily, the people who lived in that house had a phone, but it was a party line, and it was in use. The woman of the house had to tell the other people on the line to get off because she had an emergency call to make. We were soon assured that the ambulances were on their way. She also thought to call the sheriff. After she hung up the phone, she quizzed me about the wreck. I tried to tell her what little I knew.

The lady's last name was Prince. Her husband had once bought milk cows from my dad. As her face filled up with horror, she said, "Oh, dear God! I hope that wreck wasn't the Gaddis family. They just left here a few minutes ago. They buy milk from us." She wrung her hands and fingered the hem of the apron she wore. Was it a black Ford and did they have three little kids with them?"

I didn't like having to tell her, but I did. She became very agitated and started to cry. That's when I told her I needed to get back to the wreck. Upon my arrival, I was encouraged that the children were standing over by the bank, at the edge of the road. They appeared stunned, and were whimpering and holding on to each other. They looked as though they had been abandoned. The oldest little girl tried to console her siblings by telling them, "It's going to be all right now; it'll be all right." The mother franticly clopped around nearby in one flip-flop and a bare foot. She held my T-shirt over her cut eye and against the bleeding hole in her head.

She could not be consoled as she asked repeatedly about her husband's condition. I had never seen a man die before, but given how he was still motionless after all that time, I thought he had to be dead. Pete tried to console her, and kept telling her that her husband would be alright. He promised her that an ambulance was on the way. Still, she cried and prayed, "Oh, God, please let him live! Please, God, don't take him! Please, God! Please!" she screeched in a voice that cut my heart into pieces.

In the other car I could hear someone mumbling and fidgeting around. I walked over and looked in to discover in the front seat, two men whose bodies were half-way in the seat and half on the floorboard. Their bodies were twisted, and one lay on top of the other. One of the men, the one on top, was motionless. His head looked pale and was at a go-funny angle. There was a little blood coming from his mouth, and you could smell the strong odor of liquor and transmission fluid, and hot rubber hoses.

The other guy was squirming and appeared to be trying to move to the passenger side of the car. But his arm was broken, and the bone was sticking through the skin. Still, he tried to move, but without much success.

Two sheriff's cars arrived before the ambulances, and one officer asked Pete a lot of questions about what he had seen. Pete told him that he hadn't seen the wreck, but had driven up immediately afterwards. The officer wrote everything down and told Pete he would be back in touch. With that, we left with the terrible images of the wreck in our minds, and not knowing for certain who had died and who hadn't.

We picked up Jackie and returned to Isabella and Newtown by way of Harbuck, where we collected the other chicken catchers. The "Goat-man" crowd had thinned out a little as we once again passed through Five-Points.

We told the catchers about the wreck and they were all quiet after hearing the bloody details. All except for Marcus Payne, that is. He had seen plenty of blood before and wasn't affected. Marcus was bragging about a new scheme he was getting ready to pull off in just a few days. He wasn't going to catch chickens anymore. His get-rich scheme involved traveling across the country, writing checks to people in places where he had never been before.

We were late getting to the "Elbow House," but we loaded the two trucks in less than four hours. This was good time considering the feather nest was over five hundred feet long. We had to do two drive ups on the chickens.

Bill Roach was a genius at drive ups. His method was to get behind a large flock of chickens, rush at the birds while flapping his arms wildly, and shouting at the top of his voice,

"Crawford Poetry Company, Brownie Crawford, Crawford Poetry Company."

The young broilers would become horrified and run like race horses toward the front of the feather nest – up close to the chicken truck. All the lights would then be turned off at the same time, and the young chickens would sit down. His method was less effective on the older laying hens, however.

The roads were clear and we were all back home by three in the morning. We even had time to stop at Payne Gap to get a Double Cola and a candy bar. The Cola bottles were "properly deposited" while driving back up the River Road in the crisp, early morning air. Somehow or another, when the catchers would roll down the window and throw their bottles at the road signs, the middle-of-the-night moisture made them fly better. Naturally, the crew would take bets on the outcome.

I quietly removed my shoes at the entrance to Johnnie's house, took my daily bath, and tried to get a little rest before the school bus ran. But I couldn't sleep. Scenes from the wreck kept appearing in my wide-awake mind. Images of carnage and death, and the unspeakable suffering, kept rolling. The image of those dazed children and their bleeding mother with her half missing eye lid could not be erased. The odors of the wreck also lingered;

the smell of liquor mixed with hot car fluids with glassy milk. All remain burned on my brain to this day.

Pete told me a few days later that he hadn't heard from the deputies who were on the scene. But he said that he had been visited by one of the two men from the other car involved in the wreck. This was the car where the two men had smelled so strongly of liquor. The man had a cast on his arm and hobbled slowly. He told Pete that the woman's husband had been killed in the wreck, and that his brother hadn't survived, either.

Then he told Pete the purpose of his visit.

The surviving brother, who visited Pete, had been charged with vehicular homicide and was out on bond. Officers could not determine with certainty who had been the actual driver of the car at fault. Pete's visitor had come to ask Pete to testify that the dead brother had been the actual driver.

Pete reported that the man pleaded with him and, in tears, begged him. He also offered Pete money, if my brother would corroborate the testimony of the officer who had not been certain which man was the actual driver. He explained that if he was convicted of homicide, he would have to serve time, and that his aged mother would have no one to look out for her. He went on to tell Pete his lawyer believed he would most likely be sued for the injuries suffered by the wife and the children, and for the death of their father.

Pete asked the guy for a few days to consider the matter. Later, Pete told the man that he could not honor his request, that the correct man had been charged with the crime.

"I saw what I saw," were Pete's last words to the drunk driver, who had killed two men with his out-of-control car.

We both saw two men die that day. For years I worked to avoid thinking about the wreck scene. It was, and remains, too painful to willfully recall, especially the image of those vulnerable little children who lost their father that day.

An abstract and less painful image from the wreck scene recurs from time to time. It's those busted glass milk jugs. In particular, it's the lids. Although all the jugs had busted, the lids and the top part of the jugs remained solidly sealed together. Why didn't the top of the glass jugs shatter and why didn't the lids come off? All milk gushed from those jugs that day but none flowed through their normal pathway. It's weird what a teen chooses to see and remember.

17. Some Bad Luck

"The best-laid schemes o' mice an' men
gang aft agley."

Robert Burns: From "To A Mouse"

Luck was not a lady tonight and the lady's closest friend, Sister Serendipity, had jumped bail and left town. Seven rabbit's feet and a field of four leaf clovers would have changed nothing about this night. Our plans and intentions were good, I thought, but was I ever wrong! Even after the chickens had been loaded and were en route to the processing plant, bad luck would continue to dog us, to confront us, and finally leave us spent and conquered.

The night had all the potential for being easy work. We had only one truck to load, and the feather nest was located close by in Hell's Holler, Georgia. All the catchers knew this, and we all expected to be done within two hours past dark. The work load was so light that Pete was taking the night off. He had asked me to fill in for him. Picking the catchers up, taking them to the near-by work site, then returning them home would be no problem. Some of the catchers decided to also take the night off. With so few chickens to load, and so many catchers working, little money would be made.

But old Marcus didn't lay out on us. He had a baby on the way and needed the money, no matter how meager the wages. He reminded us of this when he bumped the bones. Because as he rolled the dice, he would always say, "My baby needs a pair of shoes."

Lately, Marcus had said hardly anything about his "traveling rubber check scheme." I thought the baby being on the way had maybe changed his thinking. Maybe he had thought on the possibility of doing more time in the pen. Maybe he was thinking about not being able to see his baby, if his scheme didn't work out and he got caught. But it was hard to read old Marcus. His upbeat chatter and daring facial expressions revealed little about his true frame of mind. That made him a good poker player, I thought, and some nights he made more money playing poker than he did catching chickens.

But this night wasn't just about catching chickens, because we had some plans for "extra curricular activity" after the chickens had been loaded. Those plans were enticing to some and made them decide to work, even knowing that little money would line their pockets. But some decided to work for other reasons that I would not learn about until the night was over.

Our after-work secret plan, known only to the catchers, was to drive the crate with the catchers to the "Twin City Drive-In Theater." The outside drive-in was located only about a mile from The Poetry Company, at the top of Harper Town Hill. We would attend the second showing which began about two and a half hours after dark. Driving from the feather nest, from Hell's Holler, would take only about thirty minutes. A perfectly timed plan, I thought.

I knew exactly where Hell's Holler was located. I had been there once with my dad who sometimes traded cows and mules out there. It was like most of the other rural areas surrounding the Copper Basin, with its graveled mountain roads that led to various farm houses. The area had changed some in recent years, however. Some of the farmers were now growing broilers for The Poetry Company. But what I didn't know was the name and the exact location of the broiler grower where we would work that night. From a previous experience with Brownie's directions, I knew this was most important. I went to see Brownie to get the name and the specific directions.

My brother, Brownie, was a very smart businessman, but giving directions on how to get somewhere was not his forte. His directions were always quick and included a lot of nebulous references. References like: "Drive out there past that pasture. You know. That pasture where that brindled cow with the one horn stands. Stands there and chews her cud. Turn right, right there. Don't go any farther or you'll miss it!" Invariably, the instructions included the statement, "you can't miss it." Most people who followed Brownie's directions did miss it – including the chicken truck drivers.

Brownie's directions also were usually missing at least one important detail. Late one afternoon, just before Christmas time, Brownie sent me to Chattanooga and he gave me some directions. I thought they were good directions but, they too, were lacking a critical detail. My mission was to pick up a case of Chivaz Regal Scotch that Brownie had already purchased. The fifths were Christmas gifts for his customers, as Brownie had explained before I left for the big city.

"Now hurry on over there to Chattanooga and pick that up for me. The man is there at the liquor store waiting for you now. It's that liquor store right there on Broad Street. I know you know where it is, just one block over from the Tennessee Poultry Plant on Market Street. I know you been there before. If you go to the sale barn, down there in St. Elmo, then you've gone too far. You can't miss it."

Well, Brownie hadn't told me a lie, and when I arrived on Broad Street even I was pleasantly surprised. Brownie's directions had gotten me to the right street. And sure enough, there was a liquor store right there on Broad Street… along with about twenty seven other liquor stores. They were on both the right and the left side of Broad Street. Brownie had led me to believe that only one liquor store existed on Broad Street, so I hadn't bothered to ask for the store's name. And he hadn't bothered to provide it. One by one, I stopped at each liquor store looking for the man with a case of Scotch who was waiting for me. In my long life of about fifteen years, I'd never even been in a liquor store before. The Copper

Basin area was "dry" as far as taxed liquor sales were concerned. Untaxed was another story. But after that day, I had been in eleven stores and all those stores were there on Broad Street. It was dark, and I was getting a little scared, when I finally found the right store.

Remembering that ordeal, I listened carefully to Brownie's directions for that night's work, and at first I didn't ask him any questions. I didn't want him to think I was dumb as dirt. After all, he kept saying to me, "You can't miss it."

Finally I did ask Brownie one question. "What's the feller's name where we're loading tonight?"

Brownie's comeback was finger-snap quick. "Oh, his name's Payne. He's lived out there in Hell's Holler forever. If you have trouble finding him, just stop and any of those folks who live out there will tell you where he lives."

I left The Poetry Company about an hour before dark with a plan I thought was a good way to find exactly where we were going. I picked up a crew of about ten catchers, and we headed into Hell's Holler just at dusky dark.

As we started getting into the Hell's Holler area, I noticed something peculiar about the mail boxes that lined the road. With few exceptions, the names on all the boxes read "Payne." I knew then I had a problem, because I had failed to get the first name of the "Payne" person we were trying to find. Another peculiar thing I noticed about all of these people named "Payne," was that with only a few exceptions, all the houses had a feather nest out back. So we drove on thinking that we would see the chicken truck, hoping its driver would have already found the correct "Payne."

Sure enough, there was the chicken truck, but he wasn't parked in front of a feather nest. Instead, he was simply pulled over, off to the side of the gravel road. When I pulled up beside him, Jimmy Wimpy, who was riding shotgun, asked the driver if he knew the location of the feather nest. I heard something to the affect of, "No, hell no, and these damn directions Brownie gave me ain't worth the paper they're written on." I told the driver to just sit tight; that we would go find the right location and return to lead him in.

It had been dark now for about thirty minutes when we drove away from the other truck, and I was getting a little bit worried. I decided to stop at the next house to get some better directions. When I got there, I parked the crate and, with a flashlight and Jimmy Wimpy in tow, I started walking toward the house. No sooner had we closed the truck doors than we heard a bunch of hunting dogs leap from the front porch of the house. They charged toward us barking fiercely, but we kept walking toward the house, trying to show no fear. Desperate, we tried to cajole the dogs to stop barking. As we stepped up to the front porch, the dogs followed us, sniffing us in places they shouldn't. No light was on in the house, but still I decided to knock anyhow. I didn't have much choice.

In less than a minute, the door cracked open and all I could see was two holes of a double barrel shotgun sticking out through the crack. A booming voice out of the shadows demanded, "Who are you boys and what's you want?"

I got a knot in my stomach, and my mouth turned as dry as cotton! I had never had a gun pointed at me before, and I thought for certain that I had come face to face with my mortality. But I managed to get a few words out. "Uh…Uh…My name's Joe…. Uh….Crawford and…Uh…this is Wimpey Jimmy…Uh I mean Jimmy Wimpey. Uh…Uh…We're trying to find a feller…Uh… feller named Payne…Uh…Uh…we come to load his chickens tonight."

With the gun still pointed at us, the man barked, "Hain't no Paynes lives here. Don't ye touch my dogs. Now you all get off my property."

And we did! As fast as we could, we cut a trail. We almost broke into a run. The dogs trailed behind, but soon they turned and trotted back to the house.

I was shaking so badly when I got in the crate, I gripped the steering wheel and tried to regain my nerve. Jimmy was rattled, too. He was talking a mile a minute, so fast that no one could understand what he was saying. But when they finally did understand, they all just laughed. Old Marcus couldn't stop cackling.

A change of plans was in order. I decided to first look at the name on the mailbox before I knocked on anymore doors. If it didn't read Payne, then I wasn't going there. I also decided to stop only at the Payne houses that had a light on. If I was going to be shot, I at least wanted to see it coming.

The next house had a light on and the mailbox read Payne. With flashlight in hand, I dropped from the driver's seat to the ground. From the back, I could hear old Marcus laugh sarcastically as he issued a challenge to Wimpey. "Hey, Limpey! Why don't you trot on up there to that house and knock on their door?" Marcus just kept laughing at Jimmy who was now getting some color back in his face.

It was me who knocked on the door and was rewarded by the face of a friendly man on the other side. He smiled and asked, "Where you boys been? Did you get lost?"

I explained that I had had some trouble finding the right Payne.

He just laughed. Yeah, they're more Paynes lives here in these woods than there are white folks. And I'm kin to most of them, but I don't claim them all. Not them that sells "shine," anyhow."

Once I knew we were at the right place, I explained that the truck was down the road a bit, and that we would soon get down to work. I returned to the crate and told the catchers to go with "Mr. Payne" and to do the drive up. Someone, I wasn't sure who, asked as he got out of the crate, "Has everyone laid down their guns? Is it safe to get out here? Hey, Marcus, are you kin to all these Paynes that live out here in Hell's Holler?"

Old Marcus was still ribbing Wimpey about the gunpoint incidence, but he stopped long enough to respond. "No, I got no kin out here that I know of. These Paynes are a different pack of dogs than my kin."

Bill Roach, the crew's catcher, who was also inclined to music and dance, got out of the crate and started singing a familiar church song. His harmonious and mellow pitch wafted about on the late evening air. "There'll be no pains in heaven, no pains in heaven fair. All will be glory on that day. No pains in heaven. No pains in heaven fair. No pains in heaven will be there."

In response, his older brother Bob announced, "I'm going to put a pain in your ass with my boot, if you don't stop that damn singing."

I drove back down the road to the chicken truck and the driver followed me back. The broilers had already been driven up and the catchers were waiting outside to begin the truck breakdown, when the chicken truck pulled up. We got the truck loaded in less than an hour and a half. It wasn't even a full load. That included a possum that one of the guys found scrounging around the feed storage bin. We put him into a coop by himself, and sent him on up to Knoxville with the broilers. Hambone said, as he put that possum into its temporary home, "I can see the whites of that black boy's eyes now. When he reaches in this coop and pulls out this here 'chicken' tomorrow at the plant, he's going to do a nose dive off the truck." He made the prediction with a toothless grin on his face, and his body shook with undisguised mirth.

While we were loading the chickens, the truck driver came to me and said he needed to borrow the crate for a few minutes. He said he thought he had left one of his flares on the road back down there where he had been parked, and wanted to ride back to pick it up. I thought nothing about his request and said fine.

I should have suspected something was up. Turns out he borrowed the crate to go buy some moonshine. Not for himself, but for some of the catchers. Several had given him money to buy them a pint.

I didn't know about it until we were inside the Twin City Drive In. Even then, I wasn't certain, because I had seen no one drinking. But as we pulled the crate up to the ticket station, I thought I got a whiff of it. The man selling the show tickets hadn't said anything when he stuck his head inside the crate to count us, so I figured that what I had smelled was something else. I was wrong!

The catchers didn't watch the movie, and it was a good one, I thought. "The Pink Panther" was showing, and we hadn't missed much of the beginning. But the guys soon became as unruly as a prison break crowd in a beer joint. Together they all whistled at any girl that walked by. They beat and banged on the sides and top of the crate with their fists. A couple of them got pissed with each other, got out of the crate, and were fighting right there in the graveled parking lot. Marcus convinced them to stop and get back in, by warning if they didn't, the police would be called and everyone would be arrested. Being arrested was the last thing Marcus wanted to happen. He was still on parole and risked being sent back to the pen if he so much as thought about getting into trouble.

Things calmed down a little after that, so I thought I would go to the concession stand to get a box of popcorn. When I returned, some of the catchers were gone and others were sitting outside on the truck fenders. They were still making cat-calls at all the girls and acting like they'd never seen a female before, but they were more civil than they had been earlier.

Marcus was seated in the passenger side. When he saw me, he said, "I'll be back in a few minutes. I got to go bomb Tokyo. My stomach's messed up."

In less than five minutes, one of the guys seated on the fender also took off toward the bathroom. Then the catcher on the other fender left, doing a slow lope in the same direction. Finally, the only catcher still in the back of the truck got up and left, holding his stomach and trotting fast toward the now obviously-overcrowded bathroom.

Then it hit me as to what might be going on with the crew's mad rush. I opened the glove compartment of the crate and saw immediately that something was missing. I had bought two packs of "Feenamints" earlier in the day at Tallent's Drug Store, and had placed them in the glove compartment. I hid them there because I didn't want anyone to know I had them. Feenamints were a laxative, but they looked almost identical to a type of chewing gum called "Chicklets." The two could easily be mistaken. I had bought Feenamints, albeit under other pretenses, to settle a score with a bigger, older boy at school. The guy had embarrassed me in front of a girl and I knew there was no way I could whip him in a fight. Instead, I devised a plan that involved switching the laxative for the gum, then offering him some to chew. He would get a bad case of diarrhea at the worst of all places… at school. My brilliant plan had failed, and I had a bunch of half-drunk, rowdy chicken catchers, all with a bad case of diarrhea on my hands.

In only a few minutes they all returned to stand outside the truck, huddled together. Misery was written on their faces. One of them confessed about the moonshine they had secretly been drinking, but they said nothing about taking my "chewing gum." They all concluded that the "shine" must have been bad, and they worried that they might all have been poisoned. Some even wondered aloud if they might die. They wanted to go home, so everyone loaded up, and we left "The Pink Panther" still playing on the big screen in the background. I had gotten to see very little of the movie. It took a while to get each of them home, because we had to make a few "necessary" stops along the way.

I said nothing to them about the "chewing gum" and my foiled plans. To this day, I have said nothing. If they knew then, or maybe even if they knew today, they would definitely kick my ass. It's the code. It was a bowel-churningly bad night at the theater; the drive in theater that is.

18. *Fuggeddaboutit*

"Freedom Is Not Free"
From: Korean War Veterans Memorial Wall, Washington D.C.

The religion of cleanliness was faithfully practiced the two years I stayed at my sister Johnnie's house. Only one thing excused me from having to take those daily baths, and that was already having taken one at school, after football practice. This sport was a great outlet for me, because the game not only accepted my aggressive nature, it welcomed it. Having worked with the chicken catchers, I knew no other behavior. On the football field I was as happy as a pig in mud. My team, West Fannin High, played teams from the tri-state area composed of Georgia, Tennessee, and North Carolina. Though I wasn't as big, or as fast, as my older brothers who had played the game, I was strong and plenty quick. These athletic attributes I credited to having caught chickens. Endurance, self-discipline, and most importantly, toughness, were unspoken expectations of a chicken catcher.

That toughness caused a bad misunderstanding one night in a football game we played against Murphy. I was having a great game that night. Playing defense, I had even managed to sack the quarterback a couple of times. Then it happened. What came natural for everyone, I could no longer do; I couldn't breathe. I had just sacked that quarterback, but this time something had gone wrong. He had curled his legs when we fell to the ground and his feet ended up in my stomach, under my rib cage. The air was knocked from me. No matter how hard I tried, inhaling was impossible.

I knew how the drama of an injured player on the field unfolded, and I wanted no part of this role. Gawking spectators munching on their pop-corn weren't going to be entertained by me squirming on the field, struggling for air. So, I quickly willed myself up off the ground and ran back to the forming huddle. Jamie Wright slapped my shoulder pads and congratulated me with, "Way to go Crawford."

Then the lights went out. I fell to the ground, groaning, right there in the middle of the huddle. No one but my fellow players could see me on the ground. Someone returning late to the huddle had vaguely seen Jamie's hand touch me, and asked him, "Why'd you do that for? He's on our side."

Jamie was surprised by his teammate's accusation, and protested, "I didn't do anything. Honest, all I did was tap him on his shoulder pads."

Tony Brooks, a big guy but by no measure the team's Einstein, was the last to return to the huddle. Upon seeing me lying there wiggling on the ground, he squatted down and hollered from his cavernous helmet into my face, "What's you doing down there, Crawford?"

I couldn't breathe, much less respond to Tony's perceptive inquiry.

Every man around me seemed to freeze. They just stood there, like a herd of cows, just looking down at me. Disbelief was written all over their faces. Even the referees were unaware I was down, because the huddle surrounding me blocked their view.

Coach Montgomery, from far away on the sidelines, couldn't see me either. But he ran out to the huddle when our team was flagged for a delay of game. When Coach got there, he was steaming about the penalty. But when he saw me lying there on the ground, in the center of the huddle, shock and confusion replaced anger and he immediately moved into damage control mode. He removed his cap and rubbed his brow. "Okay, which one of you morons slugged little Crawford here?" he screamed. "I know he was in good shape when he came off that last play. I saw him make that sack on the quarterback and hurry back to the huddle. Which one of you did it?....Guys, we don't have time for this horseplay. We're trying to win a game here tonight!"

They just looked at each other, clearly perplexed, but no one said anything to the coach. It was not a good time to be chatty with Coach Montgomery.

Finally, Coach lifted my midsection high off the ground and ordered, "Take a little breath, Crawford" He then lowered my midsection back to the ground. "Let it out," he barked. He repeated the sequence several times and soon, I was breathing normally on my own. I got up and jogged off the field. That was the best air I have ever breathed in my life; there, near the Hiawasse River and only about a mile from "The Miami."

Hey, Coach Montgomery, what can I say? "Thank you" isn't enough to express my gratitude for the freedom you gave me to breathe again!

My teammates were a great bunch. We had fun, even if we didn't always win. But they weren't anything like my chicken catcher buddies. These guys were children of "Ozzie and Harriet" and had it made in the shade. Unlike me, they each had a perfect home life and rarely had to struggle. They were totally ignorant of the chicken catcher culture.

Catching chickens at night and practicing football during the afternoon wasn't always possible, so I caught chickens mostly during the off season and I saved my money. My freshman year I had finally stockpiled enough to purchase a scooter made by Sears and Roebuck. It was called a "Mo-Ped," and I had picked it out from the Sears catalog. My brother, Jack, took me to Cleveland to make my purchase and haul it back. I was so excited I asked Jack if I could ride it home. He agreed and I rode it all the way from

Cleveland without stopping. That was another bad decision on my part. Top speed on the Mo-Ped was about thirty-five miles per hour. But Jack followed behind me, patiently, as I slowly made my way homeward. It was a trip that took about two hours and both my butt and legs were numb when I finally dismounted. It felt like something wet was running down my legs from my butt and, at first, I thought I was having an accident of the worst kind. But Jack knew what I was experiencing. He just laughed at me, and got a big bang out of watching me stumble around rubbing my numb butt.

Jack was a kind-hearted man and the work he did as a partner in "The Poetry Company" was not glamorous, but I am evermore convinced that his labors and his commitment were the glue that held it together. The chicken catchers and the truck drivers alike looked up to Jack and somehow saw him as their altruistic selves. He was seen as one of them instead of a part-owner of the company. He worked long hours and often these hours were during times of crises. Like driving his wrecker to a stuck chicken truck at three o'clock in the morning. He would show up and winch the trucks out of a holler, or from a muddy mountain road. His thankless efforts enabled the chicken trucks to make it to the processing plants on time. Or a crisis like rushing out to repair a chicken truck that had stopped running, usually under the worst weather conditions. Once, on an icy night, he slipped and fell under the spinning wheels of a truck. He was seriously hurt and spent about two weeks recovering in Erlanger Hospital. Sometimes it was a minor crisis like pinch-hitting for a truck driver who had been on the road for a week, and wanted to get some sleep and spend some time with his wife and family. Jack always, and without complaint, would fill in for the driver. Sometimes he even filled in for the chicken catcher foreman. He did all these things in addition to running "The Crawford Poetry Company" garage. He asked for and received no special recognition for his sacrifices.

Jack knew well about unrecognized sacrifices. He had served in the Korean War. Today, this war is called "The Forgotten War." Jack spoke little of his experience in that conflict of the late 1940s and early 1950s. But he once told me he almost froze to death over there. Another memory he shared was a painful and gory story about a work detail where he was assigned to pick up dead soldiers from the battlefield; soldiers who had been killed in combat during a winter offensive. He grimaced, looked away and, with tears forming, said, "They were frozen. All of them, and when you picked them up, it was like picking up cord wood. Their blood was frozen to their clothes and the clothes were frozen to their bodies. And they…they were frozen to the ground. Lordy, it was cold over there!"

In cold north Georgia weather, he always wore a green "army issue" coat. Winter after winter, he wore one. I sometimes wondered why he continued to wear those green army coats after he was now safe and at home, away from the war. I also wonder how people can "choose" to forget anything as tragic and painful as a war.

The Mo-Ped was great while it remained running. I was "Joe Cool" when I rode it. No one else in the Copper Basin area had one. But it stopped running one day. Jack had told me repeatedly to keep oil in the engine, but once again, I hadn't listened. The piston froze to the cylinder and the engine wall had cracked. In short, it wouldn't crank.

I thought I could fix it. All by myself, I thought I could fix it. Jack had advised me not to try, but again I didn't listen to good advice. So, like old Herbert Watkins, a fool ventured where a wise person would fear to go. I tore into that engine. Learning how an internal combustion engine works was my ostensible justification. When I had finished disassembling the small motor, there were well over fifty separate pieces. I had no clue what they were, or how they all went back together. So, in humble desperation, I went to Jack and told him of my fine mess.

Jack was sage in his advice. He didn't jump to my rescue, either. "Save all the parts. I don't have time to work on it now. But the next time I drive a chicken truck to Chattanooga, we'll load it up and take it to the Sears and Roebuck shop. They will have all the tools needed to put it back together. Maybe they can even get it running again."

About a month passed and then one day, Jack called me to tell me he would be hauling a load of chickens to Tennessee Poultry in Chattanooga the next Tuesday morning. He said to roll the Mo-Ped on over to The Poetry Company before eight o'clock, and we would load it on the chicken truck and take it to the shop. He also told me to be sure to bring all the parts. I did exactly as he asked, because I wanted to be "Wild Thing" and impress all my teenage peers once again.

On Tuesday morning I walked on over to The Poetry Company, where I learned that my brother, Jim, and the catcher Garland Brown from over at Stafford Town would be going with us. But we weren't crowded for space, because the truck was a cab over and had a sleeper. That's where Jim and I rode. Garland was going to visit his sister. The Mo-Ped and all its parts were loaded on the truck with the chickens. We had to re-arrange some chicken coops and the air gap between them, but the Mo-Ped and its parts fit nicely between the coops.

When we got to Tennessee Poultry, Jack laid it out for me. "Unload the Mo-Ped and its parts first. Park them over there by that wall and stay with them. When we get the truck in place to be unloaded, I'll cut the tractor loose from the trailer and we'll load your Mo-Ped on the tractor. We'll run bob-tail over to the repair shop."

I did as Jack said. Garland left right away to go visit with his sister with orders from my brother to be back by four o'clock that afternoon. He explained that we would be heading home at about that time.

While waiting there by the wall of the processing plant with my non-working Mo-Ped and its parts, I got an excellent idea. Why didn't I just roll the Mo-Ped on over to the repair shop right then? Maybe they could repair it right away. Maybe they would even have it fixed and we could pick it up before we left town. Besides, I was just standing there, wasting time, with nothing else to do. I explained my plan and Jim agreed to carry the parts and go with me.

We started out toward St. Elmo, because I had checked the phone book and that's where the shop was located. The phone book was old, but I figured that it didn't matter. St. Elmo turned out to be about three miles away; not too bad of a walk for young guys, but it was a bit of a slug if you were pushing a Mo-Ped or carrying a box of heavy parts. But we soldiered on toward our destination.

We got to where the shop was suppose to be about two hours later, only to learn that they had moved. The guy who rented their building wasn't sure of the new address, but he thought it was over on Twenty-Third Street. That's not so bad, I thought. And, it was back toward the poultry plant where our truck was waiting. We started out again in search of the new repair shop.

We were both whipped bad when we finally found the shop's new location. I was disappointed, however, when the manager told me that they wouldn't be able to fix my scooter that day. He gave my box of parts a skeptical look. "You may want me to give you the cost to repair this thing first. To put it back together won't cost that much. But, if we get it back together and find that something else is wrong, it may cost a lot more."

Then he pointed to the box. "Are all your parts in there?"

I told him that everything was there, but that I had broken some of them when I tried to put the engine back together. He looked at me like I was from Pluto. Then he laughed and said, "You can't put rings on a piston and put the piston in the cylinder without a ring puller. You probably broke the rings. I better get you an estimate before we start work."

Jack had been much kinder when I had told him of my attempt to repair the Mo-Ped.

I didn't know exactly what the guy was talking about, but I agreed and gave him my sister Johnnie's phone number. He said he would call in about two weeks, once he had an estimate of the total repair cost.

Feeling a little disappointed, but still relieved that I now had a plan for the repair of my Mo-Ped, Jim and I started back toward the chicken truck at Tennessee Poultry. It was less than a mile away.

We had just walked under the Twenty-Third Street Bridge when I caught the sight of two black guys walking fast from behind Jim and me. They were grown men, but one was larger than the other. The smaller one was soon walking alongside Jim, and they walked ahead of me. The larger one was walking beside me, and I started moving a little faster, looking straight ahead. After keeping pace with me for about a minute, and just looking at me without saying anything, this hulk of a man finally asked, "What's yo name, B?"

I told him "Joe," and kept walking, still looking straight ahead and refusing to make eye contact.

"Whey you stay?

"Copperhill."

Finally he asked, "You got money, white boy?"

I didn't say anything and just kept walking as fast as I could up Market Street.

That's when he reached over and grabbed my pant leg and shook it. "Yea, you got money. I kin hear it. Give me some dat money white boy, or I gonna' whup yo' ass."

Tired and irritated from the day's trials, I simply told him, "I got no money to give you."

Jim had already been asked the same question by the guy that was walking with him, and had given him some money. They seemed to be best of friends. I watched in amazement as I saw them walk along together, exchanging casual conversation.

The next thing I saw was stars spinning about my head. He had cold cocked me and I was flat on my back. I also saw in clear detail the enormous structural iron beams that supported the Twenty-Third Street Bridge. But the most unforgettable thing I next saw was a glinting steel knife blade and a large black man holding it out to his side. Then I heard him say, "I gonna' cut yo' ass, white boy. You not give me some yo' money. I gonna' cut ye."

I had a problem. Several conflicting messages were registering in my mind all at once. My personal inclination, my family raising, and my own personal pride was telling me, "Get up and fight this grown thief." But, the stars floating around my head, and the first law of nature that was being telegraphed from every part of my body, said, "If you stay here another second longer, you will be dead." I decided to retreat and fight another day. I sprang from the ground and yelled out to Jim, "Let's run, Jim!"

They were wasted words. Jim didn't need to be told because he had received the same telegraphed message from nature that I had. He was already two steps ahead of me, and gaining speed with every step. I don't know how I got by the guy's first swipe at me with that blade, but I did. Since that day I have competed in track events, have played fast-paced football, and have even been involved in a few competitive jogging events. But never in my life have I run as fast since the day that black man said to me, "I gonna cut yo' ass, white boy." I believed him.

It's a myth; Black people aren't faster runners than white people. White people simply lack proper motivation. If a knife is being swiped at your rear end by a black guy, a white boy can absolutely out-pace a black man. Resolved.

I am certain that Jim and I both set Olympic records that day. We sprinted for about a mile before we got back to the poultry plant. The two black guys followed for a while, but soon saw they were not going to catch us, so they stopped their pursuit.

When we got back to the chicken truck, I got in and locked both doors and crawled up into the sleeper. I also pulled the curtain. If I couldn't see them, they couldn't see me, I reasoned. I feared they would come looking for me and, if they found me, there was one guy who would certainly cut this white boy's ass. I stayed hid out there in the sleeper. Whereas Jim, who wasn't as fearful as I, left the truck and found Jack. He told him what had happened. In a few minutes Jack came to the truck and did his best to imitate a black street guy talking and said, "Come he uh, white boy. I want some yo' money." Then he gave it away by laughing out loud. It was that old Jack laugh – loud and genuine, and from the bottom of his big, down-home heart. I didn't even peek out from the curtains of the truck's sleeper. Then Jack said, "Come on out, Joe. They aren't going to come up here. You're safe. Get out and forget about it"

I climbed down from the truck. Things were just as Jack had said. I was safe. Anything my brother Jack said could be taken to the bank. A picture of him should be shown in the dictionary under the word "integrity." But, most revealing about my brother Jack's comments and actions, and his character on that day, was the fact that he didn't chastise me. With plenty of justification, he could have easily said, "I gave you good advice but you wouldn't listen to me." But he didn't. He truly understood and was tolerant of youthful mistakes.

The black guy who had tried to steal my money and tried to knife me stayed on my mind for a while. My true thoughts and feelings about him were not noble. These feelings were troubling and confusing to me. I was a born-again Christian and had been baptised over at Mine City Baptist Church in Ducktown. How could I possibly be a true Christian, I fretted, if I had feelings of vengeance toward those two black guys? I wanted to get a gun and go back to Chattanooga and find them both. I didn't really want to kill them, or even want to harm them. What I wanted most was for those two guys to feel the fear and humiliation and frustration they caused me. But I didn't get a gun and I didn't go back. Instead, I prayed often about my conflicted feelings and, more importantly, I tried hard this time to do as Jack had advised: I forgot about it. I had made a mistake. I had humiliated them by trespassing onto their turf. That, too, was against the chicken catcher's code. Maybe those guys knew something about fear and humiliation also, I reasoned.

The Mo-Ped was picked up later and came back home in pieces, the same way it went to Chattanooga. The cost to fix it was going to be more than the cost of a new one, so Jack brought it home on his next trip to Chattanooga. I sold it to Danny "Gomer" Oliver for twenty-five dollars.

19. Runaway Home

"I was always treated as if I had insisted on being born, in opposition to the dictates of reason, religion, and morality, and against the dissuadinig arguments of my best friends."

Charles Dickens: From Great Expectations

Garland Brown didn't return home from Chattanooga with us. I didn't immediately notice his absence. Garland was a quiet, soft spoken guy who rarely called attention to himself, but that wasn't why I hadn't missed him. My mind had been racing non-stop since the assault, with images from the Twenty-Third Street Bridge replaying in rapid fire order. The swirling stars were gone, but my face and head now throbbed. The image of that steely knife brandished by the angry black man kept flaring up. And the guilt feelings from my Olympic flight to safety gnawed at me. Like a coward, I had run away from a fight. That definitely was not up to chicken catcher code, and I guess I also feared I'd be found out.

We had just started up the River Road when Garland's absence finally penetrated my self-preoccupation. "Hey, where's Garland? I thought he was coming back with us."

Jack was void of expression and there was no tell-tale emotion in his voice, when he said, "Oh, he called to the plant, had them get me to the phone. He told me to go on home without him. Said he was going to stay with his sister for a few days."

This really didn't come as a surprise. Garland was also a whimsical sort of guy, and he played his cards close to the chest. He seemed prone to strange and unexpected appearances, and usually he showed up at the oddest of moments. For him not to be present for his return ride home was perfectly normal for Garland.

What I didn't know, and what Jack divulged to no one at the time, was the true situation regarding Garland's sister. She was being held in the downtown Chattanooga jail. Garland had traveled with us to bail her out. She was Garland's younger sister and had been missing for weeks. She was an underage run-away. Garland had told no one at first, but later confided in Jack, who advised him to contact Bud Payne, the Sheriff of Polk County. Now Garland's sister was even younger than me, but the Chattanooga police had arrested her for prostitution. The police had learned from another "lady of the street" that her home was in Polk County. The Chattanooga police informed Bud Payne of the incarceration, and Bud told Garland of his little sister's whereabouts. But there was a problem: She didn't want to go home to Stafford Town, and she had escaped from Garland. After being released on bail, the girl slipped from Garland's sight at the Trailways bus terminal. Garland stayed in Chattanooga and, on his own, located his sister again. Reluctantly she agreed to travel home with him on the bus.

Garland missed about two weeks of work because of his rescue mission. The chicken catchers didn't know where he was, and assumed that Garland himself was in jail. Some helped spread the rumor that Garland had been arrested for being drunk, for disorderly conduct, and for resisting arrest. Supposedly he was doing thirty days in the slammer down below the mountains. Pete also knew the truth of the matter, but he corrected no rumors.

That same year, my sophomore year of high school, I, too, would run away – twice to be exact. I would be a "runaway kid" once before school started and again after the school year was over. My junior year I quit school and would run away for the third time.

"Runaway," for my case was, I thought, a misnomer. Heck, I had never even heard the term "runaway kid," and didn't really know what it meant. On the farm, a runaway was a term used to identify a bawling cow that missed her lost calf, and jumped the fence to go looking for her baby. Plus, I wasn't really running away from anything the first time. It was more like I was running to something... things that were curious to me. They were things I had wondered about, and had always wanted to see. I yearned to understand them for myself, without depending on the explanations and interpretations of others.

My first runaway experience was a Huck Finn adventure. That summer I didn't stay much at Johnnie's house. Instead, I stayed on the road. I accompanied the chicken truck drivers and traveled through over thirteen states. Some were even west of the Mississippi. Traveling with the drivers, I helped them with the truck duties. Sometimes those duties included driving the truck. It gave the driver some time to sleep. This was kept confidential, and was never known by the owners of The Poetry Company. I usually earned money by helping to catch the chickens that were loaded onto the truck. I didn't make much traveling with the truck, because of my eating expenses. But, the adventure of traveling and "living off the fat of the land" was priceless. At an early age I saw places and things that teens of my age, and from my hometown, would not hope to see until years later.

For instance, I traveled to Missouri and crossed the Mississippi River on a ferry. I had been to Green Bay, Wisconsin, where I saw dairy farms with barns that looked better than most houses from back home. I drove in six-lane traffic in Chicago for the first time. Clarence Greenway, an older driver, lost his nerve and begged me to take the wheel. These trips were not without risk, but I loved them. The trek to Missouri, especially, had its potential for hazard. Three weeks prior to our crossing the Mississippi, a barge had gone under and five people drowned in the same place where we crossed the "Ol' Man." We had no problem, unless you count me having to drive the truck after we crossed the river, that is. Liquor was cheap in Missouri, so the driver bought several fifths. While driving, he opened one and took a few snorts. Before long he began to act a little crazy, like sounding off the truck's loud air horn and passing imaginary cars. He would wave to invisible people in those unseen cars as he passed them, and would tip his imaginary hat, as he visualized the chicken truck speeding past them. In the early morning hours virtually no traffic was on the long straight stretches, so his antics proved harmless. But then he begain to really scare me. After passing yet another imaginary car, he failed to return the truck to the right lane. Instead, he stayed there in that passing lane and started talking to himself. Stranger still, he talked with a strong British accent, saying stuff like "Hey mate, get that bloody bugger on the other side of the motorway....Cheerio, old chap. Come now, old chap, you must move along."

I knew the driver had never been to England and he certainly had no reason to have a British accent. It was the liquor, the Bennies and the lack of sleep talking. He had lived all his life up there on Harper Town Hill, just a few houses down from the McCaysville Methodist Church. Like most of the catchers, he hadn't been over fifty miles from home, until he started working for The Poetry Company. So I convinced the "British bloke" to pull the truck over and let me drive. I told him he should get some sleep. He continued to chug his liquor and he chatted to himself for a while, but soon he was sawing logs in the sleeper, as I gear-jammed the truck westward from the Delta.

That driver, also known as "British Guy" on the Missouri trip, was Bill Monteath. He was a former chicken catcher known by his moniker "Mon-puss." Bill had a sunken hole in the upper part of his forehead where, as a kid, he had suffered a near fatal biking accident. Bill wore his jet black hair long. A "duck tail" was his hair style, and he kept it neatly in place with "Bryl-cream." A tube of it was ever present in the truck's glove compartment. Now

Bryl-cream advertised its product on both TV and radio with the catchy jingle, "Bryl-cream…a little dab will do you"

Bill evidently never heard that commercial, because he always kept about ninety-two dabs of Bryl-cream in his artistically groomed hair. He disguised the sunken hole in his forehead by creating a curl with hair and having it sweep down over and around the beguiling indention.

Crossing the Mississippi on that ferry was one of the great thrills of my youth. The feeling was majestic and surreal. A cloud of fog hung low over the river and a ghostly, yet patriotic feeling was present. Standing outside the truck, the chickens and I breathed in the cool river air as we drifted toward the "Wild, Wild West." I imagined what it must have been like for the first settlers who made this river crossing. And, I made a personal contribution to the river in those pre-dawn hours. Earth's water cycle continues today because of my mighty contribution: I took a whiz off the aft side of the ferry into the Father of Waters.

Bill and I, and Bill's truck helper, "Pig Dillard," took a single axel trailer load of hens to Missouri. I think Pig's real name was Winston but I'm not certain. I don't remember hearing anyone ever call him by anything other than "Pig." The nickname was most appropriate. He was short and stout built; his face was rusty and freckled, and his skin looked to be covered in muddy water. His nose was flat, but had a sharp upward turn to it and, oddly, it resembled a scaled down nose of a Hampshire hog. The color and shape of his eyes also augmented his swine appearance. They were a flat, gray color and you could never really decide just what particular color they were. Often, they were bloodshot from a lack of sleep. But looking at Pig, eye-ball to eye-ball, was not possible when you talked to him. You see, Pig was cock-eyed. His right eye always seemed to be looking around the corner, while his left eye looked at the floor. But together, off to the Big Mo, we ventured that summer. Just Mon-puss, Cock-eyed Pig, me, and eighteen hundred and fifty Rhode Island Reds.

When my summer runaway adventure was over, I returned to Johnnie's to learn that a lot of things had changed during my absence. I would now be moving to another home, the place where my mom and my two older brothers, Jim and Gene, lived now. My new home was the Old Johnson Place, my brother Brownie's former residence. Brownie had bought and moved to the Joe Tipton place. Joe was Copperhill's claim to fame in the sports world; he was a professional baseball player, and was the catcher in the 1948 World Series for the Cleveland Indians. But tough times had come to Joe. He was sent back to the minors in Birmingham, and had moved his family there. So, Joe's house went up for sale. Brownie's family eventually ended up buying the famous baseball player's home.

My brother, Gene, was now a football coach and a teacher at West Fannin High School. He and Mom ran the household of my next new home. Reunited, and living with my two brothers and my Mom, I thought would be wonderful. Home again at last, I told myself. But problems lay ahead. The trident relationship with my older brother Gene would prove to be testy. In that relationship, we got to know each other better than either of us ever wanted. In my new situation, Gene was now my brother, my football coach and, by default of strange parental circumstances, my surrogate dad. My real dad had sold the farm and skipped off to Akron, Ohio, with his new bride. Gene, however, was ready and trained, and welcomed the challenge of his multi-faceted roles. He had studied Military Science while playing football at Tennessee Tech. Some say he might have been a contender in the pros, if not for his career-ending knee injury. After graduating, he could have entered the army as a second lieutenant, but he chose to coach instead. He was a Lord of Discipline, but was unprepared in handling his own, independent, free-thinking, chicken-catching brother.

At the end of my sophomore year, I left home as soon as humanly possible. I felt that I would explode if I lived one more day in my stable and over-disciplined new home setting. It was like boot camp every day. The structure of the school classroom, the football practices and now home itself, was stifling. Prisoners up at Brushy Mountain Prison had more liberty and freedom than me, I thought.

Occasionally I would run into old Carley and Charlie Tankersley, my twin chicken catcher buddies. They didn't catch chickens anymore. They had dropped out of school. Like so many who fell behind, they'd repeated grades and were now sixteen, so they didn't even enroll at high school. I ran into both of them one night down at The Show. In a crowded concession stand old Charlie saw me and unabashedly belted out, "Hey, Pissing Wolf! When did the sheriff let you out of jail? You little old shit. I ain't seen you in a coon's age. Come on over here and talk to me."

People were looking with disbelief at the loud, foul language in a public place. And Charlie did appear just a little bit abrasive, along with his bragadocious language. So, I hastened over to silence the indiscretion and visit with old Charlie.

"Where have I been? Where have you been, Stankersley?

"I've been working for the almighty dolla bill. Carlie and me, we're both working over in Dalton now. Making good money over there. A lot more than we ever made catching them damn old feather necks at night. We're staying with my sister. Her and her runt-ass husband, Burley, that is. You ought to come on over and get you a job there this summer."

About that time old Carlie swaggered up to us and Charlie punched him on the shoulder. "See Carlie! He ain't dead. You said he was probably dead by now."

Carlie replied, "I didn't say he was dead. I said he smelled like he was dead from hanging out in them feather nests so much."

Carlie and Charlie both were taller and more filled out. They were well over six feet tall. And they were dressed to kill. They were, in fact, over-dressed for The Show. But you could tell they were "put'n on the dog" and had dressed to flaunt their success. They each wore those new white Levis and a cool black shirt with its tail left hanging out. Their shirt collars were turned up to indicate they were going steady. The shirt pockets' each held a pack of "Sen-Sen." But they had put on too much of that new cologne that had just come out, the one they called "Hai Karate." But Carlie's face was scratched, and he was holding his neck tilted a bit, so I asked him, "What happened to your face there, Carlie?"

Carlie didn't respond and Charlie knew he wouldn't. With a devious expression, Charlie laughed and blurted out the dirt on his brother. "He had a run-in last Saturday night. Damn near broke his neck. But the cops didn't catch him, thank God."

"Well, why were you running from the cops?"

Charlie, still trying not to laugh and failing miserably, explained, "Well, you see, Pissing Wolf," Carlie saw a gumball machine over there in a laundry mat that he wanted. So, he went over at night, when no one could see him, planning to take it. He lifted it and was walking back to sister Pat's place, when the cops spotted him with it under his arm. They turned on their car-top flashing lights and told him to freeze, but "old shit for brains" here, he took off in a run. He made it to the neighborhood of houses where Pat lives, and was running through the back yards of those houses. He looked back to see if he had lost the cops and it happened. He didn't see it coming. He ran into a string of clotheslines that threw him on his ass. Damn near broke his neck, and scratched up his face, you see. That gumball machine was left behind. When he got to Pat's place, he swore to her that dogs had chased him and that's why he ran into the clotheslines. But he was lucky. He could be over there in the Whitfield holding cell if they had caught him."

We all laughed. Even Carlie managed to laugh a little about his misfortune. Charlie just shook his head and muttered, "Some people never learn." Then he said, "Give me a call if you decide to come over to Dalton. You ought to do it. The money's good. I got to go. I got me a sweet thang waiting for me. Is my collar still turned up?"

I indicated that their token status was intact as they both turned and swaggered away in their new shiny penny loafers.

The very next Monday I went to Dalton. I didn't tell anyone of my destination or my plans. While I wasn't certain if I would ever return home, I didn't want to be found and taken back against my will like Garland's sister. I felt rotten about not telling Mom, but I thought, "Hey, she doesn't need another person to look after. She has two other sons at home with her now." I also felt that I was a burden for everyone in my family. No one had told me that in so many words, but I felt I just didn't belong in anyone's family home. So, I had run away. Again.

In Dalton I filled out over a dozen job applications at the carpet mills. But I didn't call Charlie. I wanted to find a job by myself without anyone's help. I wore holes in my shoes walking up and down Glenwood Avenue and, near the end of the day, and at the end of Glenwood, I saw something very familiar: a chicken processing plant. The name on the sign under the big straw hat read "Red Hat Poultry." I filled out an application and was hired the very next day. One dollar and a quarter an hour was the pay rate. The plant was only a few blocks from where I'd found a place to stay. It was a single room flat, and it was right beside the Dalton Movie Theater. I walked to and from work daily.

I was strong and well conditioned from football and catching chickens, so the foreman decided to put me to work on the loading docks. Loading boxes of frozen chickens onto sealed tractor-trailers was my job. When the truck was fully loaded, we iced them down and they took off for Davenport, Iowa.

I don't know how they found me, but they did. Carlie and Charlie walked up to the loading docks, saw me standing there, waiting on the next truck to load, and they ran over to me.

"How long you been working here?" Charlie asked.

"Oh, about three weeks."

"Do you like it here?"

"I like the money. I've been working here ever day except Sundays, since I saw you all at The Show. And they pay overtime after you work over forty-eight hours. Last week I made over one hundred dollars."

"Damn! Are they hiring? We're going to put in an application. We don't make half that much over at the carpet mill. We don't get any overtime. What did you tell them to get this job?"

"They asked me what kind of work I had ever done before, and I told them I had worked on a farm and had caught chickens since I was ten years old."

Carlie and Charlie were hired. The next day, at break time, they both came out to the docks to see me.

We all stood apart from each other and just stared at each other's bright, pure white smocks. They couldn't believe how clean everything was. I broke the ice. "Where are you guys working?"

"I'm a gut puller" Charlie said. "And Carlie works over there by that big-ass drum machine, that one that slaps the feathers off the chickens."

"I've never even been inside the plant, except to walk up to the ice maker room up stairs. It's like being at the North Pole up there. Everyone complains about the ammonia, but it smells a lot better than some of those chicken houses we've worked in."

While we were standing there, a guy holding a clipboard in his hand walked up. He wore a white shirt and tie, and a pointed cap that looked like the one that a soda jerk or an ice cream vendor wears. He looked at Carlie and Charlie, but not at me, and announced, "You guys aren't supposed to be out here. It's against this plant's work rules. You need to go back to your work area."

I could tell from the look in old Charlie's eyes that he didn't like the man telling him where he should be, especially on his own time at break. Old Charlie didn't say anything at first. He just kept looking at the man's pointy white hat. Then he said, "Yes sir, mister. We were just getting ready to go. But could you tell me one thing?'

The guy in the pointy white hat seemed to swell with pride, probably from knowing that he was getting to share his vast knowledge with someone from the working class. His face wore a smirk as he looked at Charlie, who was still admiring the hat, and said, "Yes, I would be glad to tell you what I know."

Charlie asked, "What do those letters there on your hat mean. They say 'U.S.D.A.'. What does that stand for?"

The guy seemed to swell even more with pride, and he cleared his voice first, then replied, "Young man those letters stand for "United States Department of Agriculture."

Charlie's response was lickety-split quick. "Oh, I was wrong. I told Carlie here that it stood for "United States' Dumbest Ass." Then he turned to his brother, totally oblivious to the slow burn the government official was doing, and said, "Come on Carlie. Let's get back to work. He can't fire nobody. Hell, he don't even work for Red Hat. He works for the damn government. See you at lunch time, 'Pissing Wolf'. They got Salisbury Steak for lunch today at that café across the street."

The U.S.D.A. man didn't like the Tankersley twins much, but he seemed to like me. One day, when part of the plant was down for equipment repairs, he invited me to take a tour of the part of the processing complex that was still running. It was the part where the live chickens were unloaded from the chicken truck, and were hung by their legs on the conveyor lines that carried them first to a work station, where three workers were seated. That was the perfect job for me, I thought. It was a sit-down job. I wouldn't get cold, and my feet wouldn't freeze. I would be totally removed from all that ice back there on the load-out docks. Best of all, I wouldn't have to stand on my feet all day.

All that the job required was that you sit there and cut each chicken's throat as it passed by, using one of several sharp knives provided for the task. I had done this plenty of times back on the farm. I saw that when one knife got dull, you simply used another knife. You didn't even have to sharpen them; the foreman brought them to you already honed to an edge. A dream job, I thought, so I asked for a transfer and got the job.

The next morning I started my new sit-down job and at break time I felt great. Not the least bit tired and, more importantly, I was warm and my feet weren't freezing. I had started to the wash room when old Carlie and Charlie saw me. They stared at my face, busted out laughing and asked, "God damn, Pissing Wolf! What happened to you?"

I just smiled proudly and instantly got a clue to why they were laughing. My face crinkled and felt like it was cracking when I changed expressions. I dashed on down to the washroom and looked at my face that was blood red; chicken blood red. The invisible mist of chicken blood had coated my face and neck, had dried, and now it was stuck to me like plaster.

I used my entire break time trying to remove the dried blood but mostly it refused to budge. It was holding so tightly, I feared it might be there permanently. My face was sore and raw from all the scrubbing, and it reminded me of my first days at Johnnie's. Still not clean, I had to go back to my sit-down job. I skipped my next break and had no lunch. I was ashamed to wear all that blood on my face into the café across the street. But I was still proud of my easy job. I thought "How lucky am I? I don't have to stand up all day or have to pack gizzards or pull chicken guts like those other line workers." After lunch break, the foreman grinned at my discomfort and handed me a face shield. "Son, you might want to start wearing this." Ha! Ha! Funny as a crutch, I thought. I'd like to have smacked him, but instead I held it in and said nothing.

The next morning he had a more serious look on his face. He came by before the whistle blew and told me that I was no good at my new sit-down job. He said the chickens, whose throats I had cut hadn't bled out good enough. He said the USDA man recommended that I be moved, so it was back to my old job on the docks, which turned out to be my job for two more summers at Red Hat Poultry.

I returned home to stay at Mom's at the end of the first summer. Football practice was getting ready to start. I didn't like going back to the boot camp routine again, but I told myself that things would get better. Give it time. Tough it out; it was the code. At Mom's, no one made a fuss about my runaway summer. They only laughed a little about my "Bat-Man" roof-top escape from Brownie, who had tried and failed to retrieve me from my upstairs pad in Dalton. I felt that maybe it was a sign: I was finally going to be given a little breathing room. Maybe.

20. Mom Talk

"A good name is more desirable than great riches; to be esteemed is better than silver or gold."
The Book Of Proverbs, 22:1
The Holy Bible

Why did you do it Mom? Why did you give me that name?

Relieved, knowing that I was back living under her roof at the old Johnson House, Mom just smiled and said "Cobb? Well, you knew my folks' name was Cobb didn't you? I was a Cobb before I married your dad. Cobb's a good name. Did you know that my grandpaw's brother was Ty Cobb, the ball player?"

"Yes, I knew that," I replied, "but I mean the name 'Joe.' Where did that come from?" But I understood her confusion, and Mom knew I'd taken a lot of kidding at school about that middle name. They'd laughed at me and had called me "Corn-Cobb" or "Joe Cobbler." The farm boy rascals had enjoyed making my face turn red by loudly educating the city-slicker boys about a corn cob's purpose in the "out house." The name was especially embarrassing for me one school year. A new teacher and coach decided to arrange the class in alphabetical order – a temporary thing to help him learn our names. Then he went down the row and asked each student to state his or her full name as he wrote it down on the paper held in his clip-board.

"Joe Cobb Crawford" I said proudly, when it came my turn.

With a conceited smile on his face, he asked "Did you say 'Cob," like a corn-cob? How do you spell that?"

Before I could answer, both he and the class were having a good laugh about my name. My face and my oversized ears turned three shades of red, and I wanted to run out the door. But I didn't. I stayed and just looked straight ahead at the blackboard and, when the laughter stopped, I spoke up and said, "Capital C, little o, little b, little b."

Still trying not to smile he said, "Thank you, Mister Cobb. Okay. Next seat"

"William Frank Corn."

An explosion of expressions crossed his face and then, one of anger finally settled in as he looked up from his clip-board. He stared directly into Frank's blank face, and his voice took on the timbre of a storm front. "You two bozos are trying to mess with me, aren't you? Your name is Corn and his name is Cobb? Is that it?" Laughter erupted, even louder this time. "All right, you clowns get quiet right now."

Then Frank spoke up and never batted an eye. "No sir," he said quietly. "That's really my name. Frank Corn."

The laughter died quickly, and the class roll was soon mercifully completed.

I didn't correct the teacher's mistake of calling me Mr. Cobb until recess. When I did, he tried to joke with me and said, "I'm just going to call you 'Corn-Cob #1' and Corn there, I'll call him 'Corn-Cob #2'. How's that sound?"

"I'd rather be called by 'Joe' if it's okay."

He saw that I was serious about my name and said "All right, Joe." But, he called me by Joe Cobb the rest of the year, and he always would grin when he said my name. For the rest of the year, everyone who walked by Frank's desk or mine, would laugh and, like a sale barn auctioneer, would belt out in rapid fire the repeated words Corn-Cobb...Corn-Cobb...Corn-Cobb...Corn-Cobb.... My own brothers and sisters thought my middle name odd too, but not laughable. Sister Johnnie usually called me "Cobbie," unless she was serious or mad. Then she would call me Joe Cobb. I can still hear her: "Be still, Joe Cobb. I'm trying to get your rusty neck and ears clean!"

Mom would rarely talk with you about personal matters, and it seemed to make her anxious. Hard work vanquishes all problems and talk solves few of them, seemed to be her philosophy. But strangely, Mom was talking today. I let her continue.

"Well, Joe. Really it came from nowhere. We just liked the name because it was short and simple. We gave all my nine kids simple names. Never did like double names either, like 'Jim-Bob' or 'Billy-Wayne'. It takes too long to say 'em."

I persisted in my query. "Bob and Gene said I was named after that old Doc Hicks. They said his name was Joe, too. They said they remember the night I was born, and that old Doc Hicks drove down here in the freezing rain. Bob said he remembered hearing me cry, and said after I was born, old Doc Hicks went to our kitchen sink and washed up. Then he opened the oven of the wood stove and got himself some cornbread that had been baking. He told Bob that you should always eat the cornbread's crust because that's where all the nutrients are found."

Again, Mom only smiled. Then she said, "Well, all that's true I guess, except the part about the name. I think your dad may have told Doc that, and he probably thought he'd get a discount on the delivery bill. But your dad didn't give him any money. He and Doc never exchanged money on any of their trading. Your dad would always trade him a hog or a calf to pay our doctor bills, and he squared your delivery bill with Doc Hicks by giving him a white face heifer calf. You know, your dad once told Joe Williams, the veterinarian from over there at Coletown, the very same story. But you weren't named after either of those two doctors. Burt and Nolie Tipton had a boy named Joe, and that's where I first remember hearing and liking the name. You know, Joe Tipton was a great ball player. All Nolie's boys were good ball players. But Joe was the best. He moved on up and even played in the big leagues. He even played in the World Series one year. We used to listen to some of his ball games on the radio.

"But, Joe, let me tell you this. 'That old Doc Hicks' as you called him was not all bad. People talk awful about him, but he's always been there to help us whenever we needed doctoring. In fact, if it wasn't for Doc Hicks, you might not be alive today."

"Yeah, I know. He delivered me when I was born and Daddy gave him a calf. And one time he gave me a shot in the ribs where that blood sucking tick got on me and stuck. I knew that," I said smugly.

Before Mom had a chance to say anything, I went on to tell her, "Jerry Baker told me that Doc Hicks once gave his mom a shot right between the eyes? Yeah, Jerry said his mom had a bad sinus headache and she went down to that clinic, and crazy old Doc Hicks stuck her right between the eyes with that needle."

Mom, laughed a little and then, with a somber face, said, "Son, what you don't know is that if it weren't for old Doc Hicks, you wouldn't be here talking to me today. You see, you caught whooping cough when you were about three weeks old, and Doc told us that you might not live. But he found out about this new miracle drug for whooping cough, and he got some of it. Brought it down to the farm and doctored you with it. He made several trips from town down there, and you pulled out of it, and you got well. Lots of babies died from whopping cough around the time you were born. But you didn't. Old Doc Hicks is a smart man and he saved your life. You ought to speak more kindly of him."

I didn't tell Mom why I was bothered about my name, and why I didn't like Doc Hicks. But she knew it was something more than what we'd talked about. She always seemed to know – mother's intuition, I guess. I had wanted to tell her that I was glad to be home, to be living under her roof with her and my brothers again, as a family. But I didn't and I don't know why I didn't. I was wrong to deprive her of hearing those words. But she knew how I felt. She always knew how her boys felt about things before they said them.

21. Shit Happens

"Never envy the wicked! Soon they fade away like grass and disappear."

Book of Psalms, 37:1-2
The Holy Bible

Since returning home to Mom's, I had heard some terrible news: Old Marcus had been taken back to prison. Pete said it all started when Marcus stopped winning money in the poker games. Said it was like a curse from hell fell on him or something. He went on to say, "I'd lend him poker money when he'd ask for it, but then I'd tell him he ought to sit out until his luck came back. But Marcus wouldn't listen and some weeks he would lose all his wages. Some weeks he'd even lose more than his wages, and he'd end up owing me money. But he kept borrowing and wouldn't stop playing.

"Marcus would tell me, 'Pete, I got to make some fast money. I got a baby on the way. My luck's gonna' change. I know it will. If it don't, I'm gonna' go to a n***** whore house. That'll change it for sure.'"

But it didn't change, as Pete went on to explain. "One night we were going to load four trucks over there at Sand Mountain. We had sixteen guys working and the crate was packed full. About half of them were into a game of five card stud at first, including Marcus. Marcus had asked to borrow a dollar and I loaned him one, but I told him that it was the last dollar I could lend him that night. As always Marcus easily wiped out the amateurs and that left only three guys playing: Marcus, Garland, and Bill Brown. Bill wasn't losing money at stud, but for some reason he suggested that the game be changed to 'five card draw-deuces wild-no bet limit'. Marcus didn't want to change, but he lost the game vote, so the three began playing draw poker. Marcus won the first two hands. He had dealt the first hand, then Garland dealt the second hand. Bill dealt the third hand and Garland dropped out of that hand after his forth card fell, leaving him still holding a junk hand. Marcus had a pair of sevens showing and Bill only had an ace-high. But for some reason, Bill decided to bet more money than he knew Marcus had to wager. That left Marcus in a jam; he didn't have enough money to cover the bet so, Marcus yelled up front, 'Pete! Give me another dollar. I got this one in the bag. I just need a dollar to put it away.'

"I told him I couldn't do it and reminded him that I had warned him there would be no more loans that night. Marcus blew up on me and made some ugly accusations. 'You're in on this damn scam, ain't you, Pete? You and your damn cheating, bastard, cousin Brown here are trying to shake me down.'

Pete went on, "Well, old Marcus was a crook, but he wasn't a dumb crook. He was right about two of those charges he made. Bill Brown was our bastard cousin. Bill's dad was born in a whore house over in Coletown, and the father was our grandfather, Brownlow Crawford. Bill's dad had been serving prison time, but he'd gotten out a few months back, and I've been trying to work him a little. He's a good worker but I have to be careful who I work him with, because he's got one hell of a temper. You know, Joe, Bill's dad served time because he got mad and killed two men. The judge ruled it self-defense and he got off light, but you have to be careful around that man.

"Marcus was also right about the cheating. Bill had been dealing off the bottom of the deck. Marcus had grabbed the deck from Bill's hands and sure enough a 'deuce' was on the bottom of the deck. When Marcus saw the deuce, he exploded and he told Bill Brown in no uncertain terms, 'I'm going to stomp your ass, you cheating s.o.b.'

"'Don't let your battleship mouth overload your scrawny row-boat ass. I didn't put that deuce there on the bottom. It landed there on its own', Bill told him."

Pete said he pulled the crate over to the side of the road because he thought a fight was in the making and, sure enough, it was. Marcus had gone for his knife to cut Bill Brown, but Bob Roach grabbed Marcus' arm and prevented him from getting it out of his boot. "I had to put Marcus out of the crate down there around McDonald, Tennessee" Pete explained. I understood because I knew the same rule that Marcus knew: you fight, you don't ride in the crate.

Pete then told me, "Marcus didn't return home to Isabella, and he wasn't staying at the Roach's place with Bob and Bill's pregnant sister, either. No one had seen or heard from him since he was put out of the crate that night, and the TBI and Sheriff Bud Payne had been looking for him. They came to see me and I told them I had no idea where he was, but I hoped they found him. I told that TBI guy that Marcus owed me money and I needed him back to work off his debt. The TBI guy just said, 'Yeah, he owes a lot of other people in Tennessee, too. We're gonna' bring him in.'"

But like a bent dime in a poker game, old Marcus soon surfaced. He had started his check fraud scheme and had written bad checks in a string of cities that stretched from Chattanooga to Memphis. Marcus didn't make it out of the state, though, because his scheme flopped. The TBI caught up with him in Memphis. Five to ten was his sentence this time.

"He's locked up a little closer to home this time," is what Pinky, his little half-brother told me later. But, because it was his third time in, he won't even be looked at for parole for five more years."

Both families were now shattered and becoming destitute. Marcus's older brother, Melvin, was also back in the pen again, doing time for dealing in stolen property. To make matters worse, Bill Roach, Bob's younger brother, was being held down below the mountains on felonious assault charges.

Bill's fighting had landed him in a tight spot this time. A host of charges had been placed against him by the Polk County Sheriff, and they all stemmed from his drinking, fighting, and resisting an arrest. Bob said all those charges against his little brother Bill were phony, that they were really holding him in hopes Bill would rat on Melvin. They wanted Melvin's sources for the stolen goods, and thought Bill would give them up. They also thought Bill would corroborate the TBI's evidence on Marcus's check fraud plan, since things weren't too peachy between the families any longer. But they thought wrong, because they were ignorant of the code.

Bill's brother Bob was now the champion for both families. Bob was thrifty and his paltry chicken catching wages, and his cooler head were all that was holding those families together. Legal debts were mounting on both sides, and Sheriff Bud Payne's deputies kept bird-dogging Bob.

Bob's pregnant baby sister was now into her third term. She had tried to contact Marcus at prison several times, but he refused to see her or even talk to her. Bob had made a trip to the prison to visit with him and Marcus saw him, but refused to talk. All Marcus would say to him was "You should have let me kill that damn punk, Bill Brown. I'll settle the score with that bastard when I get out." When Bob mentioned the upcoming birth of his sister's baby, Marcus had demanded that the guard take him back to his cell.

Marcus didn't love the woman who was fat with his child, because he was obsessed with two other lovers: gambling and easy money. About a year earlier, he had needed money and, having failed to hock his Bulova watch to anyone in the crew, Marcus "offered" her to me for a night. Advancing him five dollars would seal the deal. He was itching to play in a planned high stakes game on the crate ride home that night. Marcus's offer embarrassed and sickened me, but not because this poor girl was a hag. She was actually very pretty, and she had a mystical quality about her. Before she got knocked up, she even had an amazing figure. That she and Marcus were ever an item had always puzzled me. Marcus was not without charm, but I couldn't understand how she had failed to see him for what he was, a self-centered scam artist. No one needed to tell me to walk away from the seduction deal, but they did. I knew the code: Stay away from another catcher's girl.

"Don't do it, Little Joe," advised Charles Cheatam's older brother Bill. "He'll bleed you for money from now on if you take a roll in the hay with that girl."

How could he do it, though? How could he offer her up to perform sex with others? She meant less to him than his last bad poker hand.

22. The Good Doctor

"I will prescribe regimens for the good of my patients according to my ability and my judgment and never do harm to anyone"

Hippocratic Oath

Marcus said you'd help me. I appreciate you doing this for me," she said ruefully as she labored to get her ballooned body into the front seat of my old '54 ford. "You know where Doc Hick's Community Clinic is over there in McCaysville, don't you?" she asked, as I reached for the car's light switch.

"Yeah, it's down below the YMCA by those railroad tracks, isn't it?"

"Yes, but don't let me out at the front door. Take me to the alley in the back of the building. Someone's going to meet me there to let me in."

"Okay," I said as I fumbled to find the car's windshield wiper switch. "How's ole Marcus doing up there?"

"Oh, he's fine," she said, hesitantly, as she looked away and watched the rain begin to fall.

But, unlike Marcus, she was no good at misrepresenting the truth. She hadn't talked to Marcus since he left town. He still was refusing to see or talk to her, or anyone from home.

As we drove toward the clinic the rain came down harder. The stench of the sulfur dioxide gas coming from the Copper Company seemed stronger than ever, and it burned both our eyes and noses. We both ignored it. Still looking away and out her car window, she broke into tears and sobbed as she said, "My baby's gonna' have it good. Gonna' have it better than me. She won't have to breathe this stinking Copper Hill air like I do. Doc said all his babies get to be rich and have things, and their daddies treat 'em good." She paused and became silent. But as she hung her head, from what sounded like the bottom of her tormented soul, I heard her moan. "God, I hate this....but I ain't got no other way. I can't do for my baby, and Marcus, he ain't never coming back." She said it as though she was talking to herself, more than to me. I said nothing and just listened and, as we drove, the only sounds heard were those of the windshield wipers and the slosh of the tires on the wet, deserted streets.

When we got to the alley she had stopped crying, but still she looked away, out through her rain-covered window. Then, as though her anguishing comments had never been said, she perked up and asked me, "Now, you can be back here Monday night at the same time, can't you?" She bravely struggled to lift herself forward, to get out of the car.

"I'll be here. Can I help you to the door?"

"No, I can do it. I'm just slow."

"Look," if you need me here before then, give me a call and I will be here."

No call ever came, and Monday evening I returned her home, without either of us saying a word.

His name was Dr. Thomas Jugarthy Hicks; in no way was it similar to mine. So, Mom had told me right, but in a sense, Bob and Gene had been correct. The name "Joe" came from Copperhill's famous baseball player whose name was "Joe Hicks Tipton." Tipton was truly named after this enigmatic doctor of the Copper Basin area, and I was named after Tipton.

Dr. Hicks was a brilliant scholar of the humanities, of medical science, and of veiled politics. He secretly ran an abortion clinic in the Copper Basin area in the early sixties. State officials from both

Georgia and Tennessee had known for years that he ran the clinic, but did nothing to stop him. He was supplying a service for which elected politicians lacked the guts or political will to confront. No politically correct substitute existed for Dr. Hick's solution to the shameful social problem of unwed, pregnant girls.

Like Caesar, Dr. Hicks was eventually taken down by members of his own profession. Their stated ostensible motivation was moral outrage. Oddly, it had taken his profession years to become outraged enough to blow the whistle on Dr. Hicks. Truth was, they loathed him for his wealth and for his easy-money medical practices. Like old Marcus, Dr. Hicks did genuinely love money – especially the easy kind.

But performing abortions wasn't where Dr. Hicks made his easy money; he made it by selling the babies of unwed, pregnant girls to childless couples who wanted to adopt a baby. Doc Hicks ran a one-stop clinic that fixed a multitude of pesky societal problems. Neither the politicians, the preachers, nor the medical profession would have to show their true colors on the issue, as long as Dr. Hicks' one-stop clinic stayed open for business.

Marcus's baby was only one of many "Hicks Babies." The baby's new father would be a wealthy doctor from Summit County, Ohio, and the baby would never want for anything. Akron, Ohio, was in Summit County and was known for its large rubber tire factories. My dad had family members who lived in Akron and, at various times, he would move there to work in the factories. When not living on the farm in Georgia, Akron was his home.

After Dad sold the farm and during the same time frame as Marcus's incarceration, Dad once again moved back to Akron. This time though, his health was failing; he had developed some kidney problems and sought the help of an Akron doctor who specialized in kidney diseases. The doctor found Dad interesting to talk with, and they quickly became trusted friends. The Akron doctor and his wife were childless and had tried unsuccessfully for years to have a baby. Dad made the doctor aware of Dr. Hicks' clinic back in Georgia and, in a matter of days, the doctor and his wife became the proud parents of Marcus's baby.

23. Well Water Thirsty

"No man is an island, entire of itself; every man is a piece of the continent, a part of the main;"
John Donne: From "No Man Is an Island"

Even though I had lived on my own, fended for myself, and feared little, living with my family gave me great comfort. The Old Johnson House, with its large back porch overlooking a green, grassy back yard shaded by tall live oak trees, was a great place to live. The trees were permanent giants of nature and of wisdom, and their canopy of green leaves covered the surroundings like an old mother hen's wings covers her biddies. Like the old house, these trees were over one hundred years old. I felt safe and protected there, and I imagined that the troubles of daily life somehow stopped at the tree line.

At that tree line, and not too far from the old house, was a large concrete object formed like a cube. It appeared to be a monument in the making, but it wasn't. Instead it was an abandoned well house now sealed and covered. The cube was about four feet tall and its perfect height and flat surface made it a good place to sit down. To just sit, do nothing and enjoy. The chicken catchers sometimes used it as a poker table, standing around it playing cards. The well was deep, and its water was good and cold, but now it wasn't needed because underground pipes from the city supplied treated water.

Early one winter evening, I was just sitting there on top of the well house with my Keds dangling, bouncing off the concrete sides of the well house. I had left the inside comfort of the Old Johnson House to get some fresh air, and to think over the day's activity. I heard a door slam and noticed that Brownie was leaving for the day. His office was only steps away from the house and, when he saw me sitting alone, he walked over to me. I noticed he was holding some papers in his hand. Only he held them out from his side and away from his body, like a chicken catcher might hold a flogging hen. When he reached where I sat, he thrust them toward me and asked, "Hey Cobbie. How would you like to owe this much money?"

I took the papers and looked at them. They were bills. A closer look and I saw some big numbers with dollar signs on them, but I didn't really understand what they meant. One thing I did understand, however, was a typed note on each of the two bills. Words in red ink read "Payment in Full Required by Month's End."

I then asked Brownie, "Is this what you owe? What do you owe them for?"

Laughing in a manner that didn't match his facial expression, Brownie glibly responded, "Chicken Feed. I owe seventy-eight thousand dollars to Pillsbury and another seventy-two thousand to Purina." He ran his hand over his face. "One hundred and fifty thousand dollars. That's lots of chicken feed, isn't it, Joe Cobb? How'd you like to owe that much money for just chicken feed?"

I looked at Brownie and said nothing. I didn't know what to say. As if it annoyed him to have to handle them again, he took back the papers and began pacing back and forth in front of the well house. He just kept looking at those bills, and would occasionally shake his head while he paced, as though he couldn't believe what he knew was reality. Finally he stopped pacing and said in a voice that was more dejected than I'd ever heard out of him, "It costs me ten cents a pound for the chicken feed needed to raise a broiler, and I can only get nine cents a pound for the broiler when they're grown and delivered to the processing plants. Those chicken houses out in Hell's Holler need feed tomorrow, but I can't order any more until I pay off one of these bills. I don't know what I'm going to do." He continued to pace, occasionally swatting the air with those bills that were obviously haunting him.

"Those chickens will starve and die. Damn bankers won't lend you any money to help you anymore, unless you can get two other people to swear and sign their names on a paper that certifies they know for a fact that I don't even need the money. I don't know, Joe. What would you do?"

I felt useless. I felt I had failed again, because I had no answer for my brother and I desperately wanted to say something that would lessen his hurt, and his anger and frustration. Was I a man or was I still a boy I wondered? Catching chickens was all I knew; I knew nothing about paying chicken feed bills or how to run a business. My big brother stopped looking at the bills and looked at me. In a consoling voice, he said, "We'll figure something out. I may need you to go with the feed truck tomorrow to see if any houses have extra feed that we can borrow to take to those growers whose feed bins are empty. Are you doing anything tomorrow?"

I just said "No, I got no plans." Still, I wished there was something I could have done sooner to help. Brownie walked over to his old Chrysler, got in and drove home. I didn't hear anymore about the chicken feed situation.

"You boys need to head on over to Brownie's. He's at home and we need you two to walk him tonight," Mom said, as the snow blew in the kitchen door we'd opened when Jim and I came bounding in. We were just getting home from school, and had shared a ride in Jim's car, a black '57 Ford. Dad had given the car to both of us, but I had sold Jim my part, because I wanted to buy my own car, a set of wheels I wouldn't have to share with anyone.

"Walk him?" I asked in a smart-ass tone. "Brownie knows how to walk. What are you talking about, Mom?"

"Your brother Brownie is …sick. He took too much medicine," Mom said, and she raised her chin, looked away, and struggled with the words.

I knew something was missing from Mom's story, but I also knew when not to try talking to her; This was one of those times. I opened the door and Jim shoved me back outside, then he followed behind me into the swirling snow.

"You're going to have to put some gas in my car" he barked, as we got back in the Ford. "You rode in my car today, and you're not riding with me anymore for free. You're a moocher, so get your money out."

"I got no money. Just let me out and I'll walk." I told him, defiance tinged my voice, almost daring him to put me out.

"I'm telling you, little brother, next time you ride in my car, you are going to buy some gas. I can't let a snot-nose like you ride for free."

This is just like old times, I thought to myself. We're fighting again; no doubt about it, I'm back home again.

Brownie and Doris's house was less than a mile away from our house. He now lived in the Joe Tipton House that had once been rented by Doctor Zackary, a local doctor who everyone called "Zack." He was their family doctor, and when we walked into the house, Zack was leaving. He looked up briefly and, like an army drill instructor with a gruff telegraphic voice, told us, "You boys see that Brownie gets a lot of exercise tonight. He needs to be kept moving so his blood will keep circulating."

Zack's words, like Mom's earlier, made no sense. But I asked no questions from "Major" Dr. Zackary who seemed to be a Joe Friday Dragnet, all-business sort of guy. Jim and I walked on back to Brownie's bedroom, where we found our brother in pajamas, seated in a straight back chair. His face looked ghostly white, almost like he had no blood in his body to keep circulating. Doris was uncharacteristically quiet and, seated on the side of their bed, was almost as pale as Brownie. She looked exhausted.

"I hope you boys are ready for a long night's walk. Whatever you do, don't let him go back to sleep." She delivered our orders in a dull, lifeless voice and left the room.

"Here, help me up." Brownie ordered in a weak voice. Jim and I got on each side of him and helped to lift him to his feet. We started to walk him. At first, we made short loops just inside his bedroom. Then we would stop to let him sit down to rest. Later, we walked him longer periods into other parts of the house. We walked him all night, only stopping for him to drink black coffee, go to the bathroom, or to take a brief rest.

Hours later and still weak, Brownie spoke. "I didn't do it. I didn't do it, you know. I just wanted to get some sleep. That's why I took those pills. That's all I wanted. I wasn't trying to kill myself. I just wanted to sleep." Then, out of the corner of his sad, tired blue eyes, he gazed at me and then at Jim, to see if we were buying it. We both looked straight ahead and kept him moving just like the doctor had ordered. When Zack returned early the next morning to check on things, we left.

24. A Smelly Business

"He has told you, O man, what is good; And what does the Lord require of you But to do justice, to love kindness, And to walk humbly with your God?"
Book of Micah, 6:8
The Holy Bible

Did he do it? Did he commit suicide? Small town, small-minded people wanted to know. As for me, I never considered it a relevant question. Instead, I pondered why civilized human beings needed an answer. I took Brownie at his word. He had said, "I just wanted to get some sleep," and he was my brother. I believed him.

In time my older brother recovered from the "big sleep." With help from his friends, he was on his feet and walked on his own. His genuine friends turned out to be people who I would never have expected. They were love in action, and though Brownie was a proud man, he came to count on them. He would continue to depend on the council of a select few, because his world would continue to crumble. The chicken feed fiasco was only the beginning of a host of calamities yet to come. They would shake him and even destroy his world, but it would never defeat him. Waves from those calamities would soon ripple through my world as well.

Monk Holden's loaded chicken truck was overweight, but he knew that, and had anticipated the heavy weigh-out number. His double axel straight truck held seventy-seven hundred and twenty, one-day-late, white-rock broilers. That was as many broilers as a larger, double-axel tractor trailer normally carried. The truck's tare weight had been higher than normal though, even when he weighed the truck in empty at the scale house at Wayne Farms in Albertville. Monk even questioned if his helper, ol' Bill Brown, had accidentally slipped back into the truck when the empty weight was being taken? Too late to do anything about that now, he reasoned and, if a mistake had been made, the error would benefit The Poetry Company, since Bill didn't weigh out with the loaded truck – he rode home with the crew instead. Brownie had bought the "stranded" chickens for six cents a pound and would sell them the next day to Harrison Poultry in Knoxville, Tennessee, for double that price.

But now Monk had a bigger problem: How would he get these feather necks to Knoxville on time, without the ICC or the Tennessee D.O.T. shutting him down? The ICC and the Tennessee D.O.T. had a weight limit restriction of thirty-four thousand pounds for a tandem axel truck, and his truck's weigh-out had been fifty-nine thousand and forty pounds. Monk's solution was to stay off the main highways, especially those he knew to have truck weigh-in stations, where his truck would be weighed and frozen in place until some load was removed. That would mean by-passing Chattanooga and a few other of the larger towns, that he normally traveled through, when delivering chickens from the Sand Mountain, Alabama, area to Knoxville.

The scales by-pass plan worked great until Monk reached the Rockwood, Tennessee, area. The area was mountainous, and had large gorges and a river that were spanned by steel truss bridges. Signs were posted at these bridges to warn drivers of the structure's weight limit, but Monk knew that the spans could handle more than their posted weight limit. He had often seen large farm vehicles that weighed far more than his chicken truck pass over them safely. So, Monk drove on, slowing down only a little, as he passed over bridge after bridge on his way to Knoxville.

But then it happened; something Monk hadn't considered. The truck's height became an unexpected and immediate problem.

Monk saw it coming, but it was too late to take corrective action. A cross brace dangled from the top of the bridge and instantly, like a javelin, the brace pierced the top of the chicken truck, grabbed it, and pivoted the truck's motion toward the side of the bridge and off into the gorge and the river bottom that lay two hundred feet below. Worse yet, the truck's gas tank had been ruptured and gasoline rained down onto the rocks and into the river below, as the truck plummeted. An explosion echoed through the gorge and was heard three miles away, when the truck struck the castle-like rocks and the river bed below.

Three coon hunters in the area had heard the explosion. They raced to their pick-up truck and drove to the bridge, where orange and blue flames billowed up from the inferno below. They were not immediately able to tell what was burning, but reported to the Tennessee State Patrol, "Smells like somebody burnt Sunday Dinner down yonder."

Monk miraculously escaped from the burning truck, but was badly burned, and would be scarred for life. Bill Brown had dealt himself another lucky hand in deciding to ride home in the crate, instead of traveling with Monk to Knoxville. That smoldering "Pit" down yonder at the bottom of that deep Tennessee gorge didn't just scar Monk's body. The horror of that accident was burnt on his brain. He suffered greatly but, in time, he fought his way through to the other side. It was the code of the chicken catcher: Never drop your chickens until they're put in their coop!

25. Sweet Sorrow

"There was a child went forth every day; And the first object he look'd upon, that object he became,
And that object became part of him for the day or a certain part of the day,
Or for many years or stretching cycles of years."
Walt Whitman: There Was a Child Went Forth

And everyone lived happily ever after with their days filled with nothing but rainbows and butterflies. NOT. Life went on, but not without struggles, disappointments, and pain. Monk Holden had survived the Tennessee bridge's free fall, but lost his wife to another man, just as Carl had lost her to him. Monk was later shot and killed in a domestic squabble by the next husband of his lost wife.

Carlie Tankersley, the twin of Charlie died young. His motto of "Live fast and die young" was realized. He grew up, slipped away from Charlie's shelter, but continued to take risks. Carlie loved his nice clothes and had always talked of having a fast car. Ironically, he lost his life in a car wreck.

Jimmy Wimpey also saw his share of vicisitudes. He grew up and got a good construction job, but nearly lost his life in an electrical accident.

Herbert Watkins moved away with his wife to Texas to work for an earth moving company. His younger brother Bennie, moved to Calhoun, Georgia, drove a truck but died early of a heart attack. T-Bone, stayed in the Fannin County, Georgia area. High on my Bucket List is to find T-Bone and the other chicken catchers to recollect those shared memories of our youth.

Many of the catchers I will probably never see again. Time has taken them on to the big drive up on the other side. But time will never take away my treasured memories of them and those green age days we shared on our climb up the rough side of the mountain.

Brownie and Jack lost The Poetry Company. A tsunami wave of outside economics took them under. To say that "they worked hard to keep the Crawford Poultry Company afloat" so grossly understates the real truth of the situation. The better part of their lives' energy had been invested in that company. Many jobless people of the Great Copper Basin were provided a ticket to board the progress train because of their drive, their vision and their compassion. But like the soil of the eroding red hills of the Great Copper Basin, they were washed away. They and their families moved to Mississippi, where they started over again from scratch, and they thrived. A city sports park named to honor and commemorate their contribution proudly stands today in their new hometown. Poetically fitting, it's located right smack dab in the middle of William Faulkner country – Yalabusha County, Water Valley, Mississippi. They, too, have gone on to the big drive up on the other side. God, I miss my brothers so much!

As for "Little Joe," well I did okay, too. I even had my own chicken catching crew. During my junior year of high school, some of the boys on the football team and I caught a few feather necks at night. It kept the team in good physical condition. I was elected the class president that year and was probably the only class president ever known to quit school and run away from home. Now, I had one thing I feared worse than needles, and I just couldn't gather the nerve to overcome it – standing before a crowd and making a speech. Spending three days in The Pit would be far easier than making a speech, I thought. But, as the class president, it was incumbent upon me to give a speech at the Senior Prom, a school gala event. Like me, most boys hated attending this annual event while most girls adored and dreamed all year of being there. I felt about as qualified for this challenge as a one-legged man would be at an ass-kicking competition. So, I ran again.

This time I ran and hid out in Clearwater, Florida. I got a job and stayed in a boarding house where mostly old, indigent people stayed. For two weeks I stayed there. One day, as I drove to my minimum wage job, I asked myself, "Is this what you want to do for the rest of your life?" It really wasn't a question and, I reasoned again, "Maybe education is not so bad." So, I loaded up the old '54 Ford and headed back north. As I traveled, I tried to get myself psyched-up for the dreaded speech that awaited me. Luckily, the prom sponsors had decided I would never return to school again, so they asked the class vice-president to make the speech. Jane Godfrey was a princess, looked the part, and she handled the speech magnificently for the upperclassmen's social event of the year.

In the beginning of my senior year of school I married the love of my life. She still is. To love and be loved changed me – Joe, the chicken catcher. My sister, Doris, the poet of the family, had once tried to tell me this: love gives meaning to life, gives you a fresh new view, while moving you to a better place – like riding up front, shotgun, in the crate.

I got an afternoon job sacking groceries at the Home Store Grocery. Love and marriage changed my perspective. For the first time in my life I took studying seriously. As a married high school student, I thought my football days were over, but once more I was wrong. My brothers again came to the rescue. This time it was my brother, "General Gene," who threw out the life line. Gene wanted me back on the football team. He approached Brownie and Jack and encouraged them to give me a job working in their chicken hatchery on Saturdays and Sundays. It was the first time in my life that I didn't have to catch chickens at night. Some nights I'd wake up and feel like I was in the wrong place. Had I been misplaced? Why was I so comfortable? I had a strange guilty and drawn feeling that I should be in a chicken house somewhere. Maybe up around Hangin Dog, North Carolina, or over in the "Elbow House" in Ringgold, Georgia.

Again I played football. A little awkward and embarrassing at first, being a married man playing on a football team with teenage boys, but somehow it worked. No, we weren't state champs, but we played respectable ball. We were small, quick and few in number, and most of us had to play both sides of the football. No team was better conditioned than the mighty Yellow Jackets of Fannin County. I could kick the football well and, for the first three weeks, I held the state record for kickoff distance. All my kicks went into the end zone. Not a big deal by today's standards, but those were pre-soccer style days.

After graduation my options were limited: Go to college or go to The University of Viet Nam. I chose the former, but I had a small problem: no funds to pay for college. Again my brothers came to the rescue. Brother Bob loaned me three hundred dollars and got me a job installing elevators in Charlotte, North Carolina. I worked during the days and went to night school, and became a pretty good student. Mom was especially proud of me when I received my Bachelors with honors. My other siblings, some of whom Dad had made quit school to work on the farm, also thought it cool as well.

Brother Bob, the reputed miser of all my siblings, refused to be repaid the loan and encouraged me to continue my education further. I took his advice. I owe Bob so much more than just three hundred dollars.

How did I make it through my age of adolescence? It was the Code of the Chicken Catcher that saw me safely through to the other side. God does truly work in mysterious ways.

Epilogue

It's been half a century, give or take, since those days when the Crawford Poultry Company was one of the two major employers in the Great Copper Basin... when Marcus Payne and Bill Roach were such an integral part of my life. I'll never be able to forget T-Bone Watkins and Randal Cearley. That goes double for the Tankersley Twins, no pun intended! Each of my chicken catcher buddies were unique people, and the lessons I learned from them were probably premature for a boy my age, but nurturing nonetheless. I can honestly say I wouldn't be where I am today, had it not been for their influences.

Are the memories of these men and those situations the total story? The quick and short answer would be, "NO!" My internal hard drive overflows with still more recollections and life-lessons. Perhaps, at a time to come, I'll share more about these principally un-learned, yet highly astute individuals, who were my role models, my mentors and my inspiration. Even through their negative actions, I acquired positive validation.

It was almost like poetry in motion… the Crawford Poetry Company, that is!

About the Author

Joe Cobb Crawford, PE, is a native of The Great Copper Basin of Tennessee, the setting and material source for his first coming-of-age novel. A hampered student who quit school three times, he married at age seventeen, but became a scholar and graduated with honors from Southern Technical Institute in Marietta, Georgia, and holds a MS degree in Engineering from the University of Tennessee in Knoxville, Tennessee. Joc is a licensed professional engineer in numerous southeastern states. The Poultry Industry, of which he writes, is only one of many industries where today he trains people in electrical engineering, safety, and energy conservation initiatives. He is a Senior Member of the Institute of Electrical and Electronics Engineers, and he works with the IEEE to develop electrical safety design standards. He resides on Lake Hartwell near Toccoa, Georgia, with his wife of forty-three years, Susan, an elementary school principal. Joe credits faith, family and early lessons learned from the most unlikely souls of God's humanity, his fellow "Chicken Catchers," as the inspiration for his life's accomplishments, and the well-head from which his stories are drawn.

Joe Cobb Crawford

Goudy Old Style
Type and design by Karen P. Stone